Just Doing My Job

Just Doing My Job

COPS, FIRIES, AMBOS

Everyday Australians with extraordinary stories

JAMES KNIGHT

HACHETTE AUSTRALIA

Cover photograph of Queensland Fireman
Brown with a young bus accident victim.
The urgency shows on the firefighters face.
Courtesy Newspix/Chris Higgins.

HACHETTE AUSTRALIA

First published in Australia and New Zealand in 2006
by Hodder Australia
This edition published by Hachette Australia in 2007
(An imprint of Hachette Livre Australia Pty Limited)
Level 17, 207 Kent Street, Sydney NSW 2000
Website: www.hachette.com.au

National Library of Australia
Cataloguing-in-Publication data

Knight, James, 1967- .
Just doing my job : everyday Australians with extraordinary stories.

ISBN 978 0 7336 2195 6 (pbk.).

1. Police - Australia - Anecdotes. 2. Emergency medical technicians - Australia - Anecdotes. 3. Fire fighters - Australia - Anecdotes. 4. Ambulance drivers - Australia - Anecdotes. I. Title.

362.180994

Text design and typesetting by Bookhouse, Sydney
Cover design and picture sections by Luke Causby/Blue Cork
Printed in Australia by Griffin Press, Adelaide

Hachette Livre Australia's policy is to use papers that are natural, renewable and recyclable products and made from wood grown in sustainable forests. The logging and manufacturing processes are expected to conform to the environmental regulations of the country of origin.

To all the Cops, Firies and Ambos across Australia.

Thank you.

And to Clare.

You're beautiful. I love you.

CONTENTS

PROLOGUE

Many of us have been in this position: we are driving along when we hear a siren, we look into the rear-vision mirror and see the flashing lights of an emergency vehicle weaving through the traffic, we pull over and let it pass, and give only a fleeting thought to where the vehicle is heading. A car accident? A cardiac arrest? A robbery? A gas explosion? In that brief moment we may pity the victim . . . but how often do we think of the people on their way to the emergency? The police, the firefighters, the ambulance officers.

They are regularly called out to incidents that most of us (luckily) will rarely, if ever, experience in our entire lives.

Imagine what it must be like to attend a fatal accident in which you know the victim; or to pull a dying baby out of a fire who is dressed in the same style of pyjamas that your child wears; or to have to knock on a stranger's door and tell a woman that her husband has been murdered, while in the background her children are preparing for a slumber party. How would you feel? Then again, you might have to chase a naked man through a five-star resort; or be escorted through a seedy establishment by a six-foot-four transvestite; or have to assist an inebriated woman who has somehow fallen into a shopping trolley and become stuck! In the course of one shift, you may be shot at, spat at, punched, verbally abused, or hugged and cheered . . . The diversity of situations and emotions is surely as great as any person in any line of work ever experiences.

Cops, firies and ambos are at the front line of our society and so often we see only the uniform. If, however, we look behind both the front line and the uniforms, we discover some remarkable people who've endured their own hardships and challenges and yet still step up to help others.

This book brings together some of these people. The stories in the following pages are snapshots of lives and events that aim to give the general public a better understanding of the individuals who serve our community.

Just Doing My Job is about a group of everyday people who not only have their occupations in common, but who all reflect the richness and strength of the human spirit.

While doing the interviews for this book, I was told time and again by the subjects: 'I hope you find my story interesting. I'm nothing special.'

But, as you'll discover, each person most certainly is.

MARK BURNS

'Home Truths'

'I'm a better person for doing this job simply because it's given me a better outlook on the community as a whole, and a better sense of self worth.'

Mark Burns, MICA Paramedic, Rural Ambulance Victoria

The air is filled with dust and the sharp citrus scent of the Mallee. It is hot, pushing 40 degrees. The paramedics wind down their windows and drive quietly out of town. It's an unwritten rule that the siren isn't used after 11 pm unless it's absolutely necessary. People are sleeping; let them sleep. One paramedic, with hands on wheel and head up, follows the road's straight whites and twists of yellow; the other, with hands holding a clipboard and head down, does the paperwork. There is little noise, except for the rumble of a diesel engine and the *tick*, *tick*, *tick* of the lights turning on the roof. Outside, the country flashes by in shades of emergency red and blue. Mark Burns knows this journey well.

'You're awake, your partner is awake, but no one else is,' he says. 'Everyone else is tucked up in bed under the sheets with a fan blowing, and you're out in the middle of nowhere going to a job. It can be pretty lonely. You feel like a ghost getting around. Then morning comes, we go home, the rest of the town gets up, and no one knows what might have happened the night before. It's a weird feeling.'

When Mark was a boy, he ran away from his home and family on Christmas Day. Whatever his reason—he thinks it may have been prompted by a gift he didn't want—his desire to leave behind the world that he knew lasted a mere few hundred metres. Once reaching the bus stop at the corner of the street, he sat down on a bench and waited for one of his sisters to find him. In the 30 years since then, he has walked out countless front doors, and travelled endless kilometres. Yet, he has never left home.

Mark lives in Swan Hill, the town in which he was born and raised. He was the youngest of nine children to Bob and Betty of Bath Street, two well-respected community figures. Bob was an accountant, and for many years he served on the Board at the local hospital where his wife worked as a nurse.

With a population of about 10,000, Swan Hill is a busy service centre for outlying rural areas in north-western Victoria, 350 kilometres from Melbourne. The Murray River winds past its northern edge, and when the water is low it's possible to skim a stone or punt a Sherrin into New South Wales. In this area with a majestic natural border, Mark has pursued a career which doesn't allow for barriers. Paramedics go wherever they have to. However, for a man who knows the lay of the land so well, there is a painful obstacle that must be overcome; in his job it's inevitable that sometimes he must attend to a person he knows. Mark leans forward in the chair at his kitchen table, he rubs his hand across his goatee beard. He thinks carefully before he speaks.

'The difference between the officers in the metropolitan areas and the country areas is that the metro guys know that their next job is only ten minutes away. They get dispatched to one job after another. You still see what you see and hear what you hear, but it's not as personal. You can go home and leave those people behind. But here, there are the constant reminders. You might see a deceased person's parents the next day, or you might pass their house. There's always something that can take you back.'

Since joining Rural Ambulance Victoria in 1992, Mark has been trapped in the middle of someone else's grief many times. At dams, pools, homes, roadsides . . . nowhere is off-limits to tragedy. Although paramedics commonly try to distance themselves from emotions, it is awful just to picture how difficult that must be when there is a personal link with the victim or family.

Just try to imagine what it must be like. You walk into a family backyard of someone you know. A child is lying on the ground with serious injuries. Other family members are howling. You focus on the patient in front of you. The sounds are deafening. Somehow you have to concentrate. You put an IV line into the child. You see a young policewoman next to you. You ask her to help you. She has tears streaming down her face. For the briefest of moments you want to yell: 'This is bullshit! Why am I here? Why am I responsible for this?' The howls and screams seem louder. Another second passes and you want to yell: 'Shut up! Everyone just shut up and let me get on with what I'm meant to do.' You eventually take the child to hospital. The family is with you in the resuscitation room. You want to find a trapdoor, any way out. There are police officers, nurses, doctors, and those awful screams. You try to avoid eye contact with anyone in the family. You move away as the doctors and nurses take over. It's time to do the paperwork. It would be easier if you were a smoker. Then you would have an excuse to go outside. But you stay close by with a clipboard, a pen, and your thoughts.

The child dies.

The very next day you are back in the hospital, wheeling in a teenage car accident victim. His father, whom you don't know, grabs you by the shirt, stares at you and demands an answer: 'Is he going to live? Don't bullshit me!' Then his wife walks in. You recognise her. She is crying. What can you do to console her when you know her son is going to die?

This has happened to Mark Burns. It's a part of the job that he frankly admits can leave scars. He says, 'The hardest part is facing the relatives and friends of a deceased person. All you want to do is shrink into the background and get out of there. And you certainly don't want to be the one to tell them that their son, daughter, mother, closest friend is deceased. And when you actually know them, which is going to happen a bit in a small community, it's only human nature that there's going to be some parts that stick with you.'

Mark has one incident that sticks with him more than any other. He doesn't wish it to be written about here—his thoughts are still too raw, and there is no need to again invade the lives of people who have suffered so much. Vivid recollections of this incident haunt Mark at the strangest times. They come without warning, and leave just as suddenly. He says with painful frankness, 'I live and work with people who are constant reminders of the tragedies we face. There is an understanding between fellow emergency workers, a knowing look between two people that says a thousand words. There is a strange silence. At times you smell certain things and you hear sounds that bring all these things back. You know all those ads on TV about crime, rescue, emergency, hospital shows? I can't watch them. Strange huh? I can go out and see all that first hand, and do my best at that time. I intubate, put IV lines in, place lung-inflating needles strategically into the chests of the sick and dying. That is real life but I can't watch that on an ad on television.

'Any professional person who says things don't worry them is a bullshit artist. They do. They bloody do! There's nothing, no magical sentence that anyone can say that can make you feel any better. We have great professional counselling teams, and crisis management, but for me only time will help.

'We had two other serious accidents in a row just before that, and I had times where I thought "Jesus, what am I doing? Do I need this job?" I gave really serious consideration to throwing it in. But then I went back to work and started helping people again. I helped a child really sick with asthma, and I could see the parents thinking: "Well, how did you do that?" That type of job made it all worthwhile again, and I thought the world wasn't such a bad place after all. And that's why I do it. I want to help this community. You just have to accept there are going to be those days when you're going to ask: "Why did this happen?"'

While Mark had few reasons to pose that question early in his career, in 1997 he was involved in an incident that ensured he would always have a sensitive understanding of distressed parents and their relationships with emergency medical staff. The case concerned young baby William. He was born two weeks premature after his mother, Alison, had suffered high blood pressure and a few other irregularities late in her pregnancy. Mark was close by when the baby was born in

Swan Hill Hospital. It was very late at night, and the excited father restrained himself from disturbing all of his friends with the news. Instead, he limited his calls to the mates he knew would still be awake: a long-haul truck driver, and a group of work colleagues: 'Hey guess what! It's a boy! It's healthy and Alison is fine.'

But within hours, William started having seizures. The doctors were puzzled, the mother was anxious, and the nervous father wanted answers. It was promptly decided that William would be flown by plane in the Neo Natal Emergency Transport Service (NNETS) to the Royal Women's Hospital. Mark had seen all this happen before. NNETS was a most professional service, and although the parents had every reason to worry, they could be grateful that their baby was in the best possible hands. But this time was different. This time Mark was the parent.

As the NNETS plane took off, Mark and Alison prepared for a four-and-a-half-hour drive. Every moment away from their son was time that could never be recovered. They'd been childhood sweethearts, married for eight years and now they'd just entered the radiant age of parenthood, but it wasn't meant to start like this.

By the time they arrived at the hospital, William was still having fits. He was in Intensive Care, being treated by some of the best medical specialists in Australia. Mark knew better than most that all he and Alison could do was wait. William's tiny, vulnerable body twisted and jerked for 24 hours and then, without explanation, the seizures became less frequent. Relief followed, as parents and doctors both watched the patient improve gradually. After twelve days in Intensive Care William was released from hospital, and Mark returned to Swan Hill not only with his wife and first child, but a greater empathy for everyone involved in such dramas. He says: 'To this day we still don't know why it happened. When we were in the middle of it, all I wanted was answers. I see it so many times with people now when I'm working. All they want is for me to tell them what is wrong, and what's going to happen. Now I can really understand what they go through, and why they go through it. I can understand when they get emotional and distressed, and I understand the pressures the people around them are under. You're going to get all sorts of reactions when you are dealing with life. William gave me a true appreciation of that. I've now seen it from both sides, and that has given me a greater reason to think that I'm doing the right job.'

William's illness forced Mark to postpone his training to become a Mobile Intensive Care (MICA) Officer, the highest level of paramedic in Victoria. He eventually started the course the following year, and after a demanding schedule which included a series of exams, interviews, and 24 weeks on the road with a clinical instructor, he graduated. MICA officers have incredible responsibilities. Put simply, they can perform everything that happens in a hospital's emergency department, other than surgery. Nowadays Mark is also a MICA Clinical Instructor.

Having such life-saving knowledge and experience is far removed from where Mark started his working life as a sixteen year old selling timber in a hardware store. After that he spent nine years as a carpenter, during which time he completed a first aid course which initially provoked his interest in a possible career change. The course was conducted one night a week for nearly two months at Swan Hill Ambulance Station. The instructor was a senior paramedic whose stories fascinated Mark to the extent that his questions soon spread beyond:

'What do you do if a person has collapsed?' to 'How did you get into this job? What qualifications do you need?'

Three years and at least as many rejections later, he was finally accepted. After his very first day at the Ambulance Officers' Training Centre in Melbourne, Mark rang Alison and said: 'I don't know what I'm doing here. This is just too hard!' Although his mother had been a nurse, he had no real medical background. He was unsettled by the lectures about cells and tissues, the types of subjects that he'd paid scant regard to when at school. But as time progressed, he discovered that diagrams of a mitochondria or an endoplasmic reticulum were stepping stones to intriguing discussions that ranged from the resuscitation of babies to handling dangerous psychiatric patients. Mark was soon enjoying it.

When he started on the road, his nerves and expectations were dulled by his first entry into the Swan Hill Ambulance Station as an employee. He anticipated a buzz of activity but instead he received an invitation to play cards. However, his thoughts of 'how boring is this?' were swamped in the following weeks by the sight of the buttons on the phone exchange lighting up, and the announcements over the loudspeakers that came afterwards:

'Officers Burns and Smith* required to attend a . . .'

Mark recalls: 'In those early days, as soon as I heard my name, my heart would sink to my stomach and I'd think, "Oh God I've got to go and do a job now!" I was comfortable with all the theory, but experience and confidence was another thing. I was on tenterhooks all the time. It took a good twelve months to overcome that. I was with guys who'd been there for fifteen, twenty years. They could tell stories of some horrendous things that happened. I used to think "Am I ever going to see that sort of stuff, and if I do, I hope it's not today!"'

Those sights inevitably came. His first motor vehicle accident call-out started with a curious power blackout at the ambulance station. The phone rang in the darkness, prompting one of the officers to suggest: 'This'll be a prang somewhere.'

It was. Mark and his partner soon arrived at the scene of a car that had been torn in two after it had clipped a gutter, launched into the air, and slammed into a power pole, thus affecting electricity supplies to some parts of town. While the engine had skidded some distance away, the cabin of the car had remained intact. Two young men were screaming inside it.

Before stepping out of the ambulance Mark took a deep breath and quietly told himself: 'I am actually part of this. I can't back out now. This is what I'm here for. I am prepared.'

Quite miraculously, neither of the accident victims was seriously hurt. After needing to get into the car to attend them, Mark gained confidence from knowing that his patients didn't realise he was a rookie. He treated them, reassured them, and by the time he was back at the station, he gave himself a quiet pat on the back. Yes, he had indeed been prepared.

But was he ready for the ugliest part of his job? Thankfully time guided him carefully for more than two years before he needed to face it. The moment happened when he was called to an accident site at which a person was slumped in the front seat of a car that had hit a tree at high speed. With no access possible through the vehicle's front, Mark climbed into the back and tried to lift up the victim's head. The injuries were gruesome, and Mark was holding death in his hands for the first time.

'You can never be totally prepared, but it helped that I had a fair bit of experience behind me by the time that happened,' he says. 'To this day I can't explain what it's like to be with a dead person. There is an eerie silence but there is a presence. You almost expect them to say something to you. It's really, really strange. It doesn't matter if it's a four-year-old child or an eighty-year-old grandmother, it's still the same.'

In a job that surely offers as diverse a range of experiences as any on the planet, Mark has also held the freshest of life in his hands. He has been involved professionally in two births, one of which was in the back of the ambulance in the 'middle of nowhere' between Swan Hill and the major regional centre, Bendigo, 180 kilometres away. Once the baby was born, the call went through to Ambulance Control: 'We now have an extra passenger on board.' The cheers crackled over the radio. Mark acknowledges these are the moments that help soften the impact of the hard times.

'It's a powerful feeling to think that you're involved in bringing a new life into the world. Like the fatalities you go to, it's the emotions of the people around you that affect you. You have the father in the back of the ambulance with you, he's smiling from ear to ear, and he just can't stop patting you on the back and thanking you. It's a great feeling.'

Between the poles of life and death are the ordinary, bizarre, unexpected, and predictable moments that ensure Mark's next job will never be the same as his last. From the man who collapses in a fit after snorting cocaine, to the inebriated woman who falls into a shopping trolley and gets stuck. From the diabetic who needs insulin, to the asthma sufferer who needs air. There are those too young to understand what is happening. There are those so old that their memory can walk the main street as if it was 70 years ago. There are 'the frequent flyers', the ones who often need nothing more than a conversation and a cup of tea. There are the first-timers whose nerves are so brittle that they scream whenever they're touched. There are car accidents, domestic violence cases, drownings, football tackles, cardiac arrests, drunken knuckle-ups, horse falls. There are broken legs, fractured cheeks, bloodied noses, shattered teeth, allergies, dislocations, cracked skulls . . .

None of us can imagine what it must be like unless we have been there.

And Mark Burns has.

When growing up, he thought he might be an 'economic genius', but instead, he left school and built houses. Nowadays, the foundations of his life are people. He is 39 years old. He is a husband, and father of three children (twins Max and Anna were born in 2000). He also has a bigger family, some 40,000 people across the Swan Hill district. The very nature of his job means the reunions he attends can cut to the core of human emotions. Being a MICA paramedic in a country community is a difficult, often heart-wrenching job. Trauma and sadness are simply occupational hazards. Mark accepts this. Despite the tragedies he faces, he knows he belongs in the job. It's a belonging that comes with living an entire life in the one place. Born in Swan Hill Hospital, he has grown into a man who understands he was born to serve his community.

'After all,' he says, 'this is my town, and I have to give something to it.'

*Smith is an invented name.

Author's note: In Victoria, base level ambulance officers are referred to as paramedics.

BERNIE AUST

'Intersections'

'The main intersections of your life determine who you are. I have been so lucky.'

Chief Superintendent Bernard Aust, New South Wales Police Service

It was about 8 pm on a clear and mild spring evening. Probationary Constable Bernard (Bernie) Aust was on relief guard duty at the Consulate General of Lebanon in Sydney's eastern suburbs. While his colleagues enjoyed a dinner break, Bernie walked along a dimly lit driveway at the back of the Consulate, checking the perimeter of the premises. The 23-year-old PC who'd been a policeman for just four weeks hoped such boring work would soon end, he was keen to experience the real adventures of being a law enforcer, but for the moment he could only count his steps and focus on the job at hand. After deciding all was safe and secure at the rear of the building, he turned around to walk back the way he'd come. As he did, he saw a man standing in the shadows about five metres ahead of him. Although Bernie couldn't see properly in the weak light, he could tell the stranger was holding a long, thin object, like a fishing rod.

'Stop! What are you doing? Who are you?'

The man didn't speak. He moved closer, and Bernie suddenly realised the danger he was in. The rod was in fact a single-barrelled shotgun, and it was pointed at his stomach.

Probationary Constable Bernie Aust was about to begin a fight for his life. It was Friday 13 October 1972—Black Friday, a fitting day for a frightening prediction to come true.

***

If time could stand still perhaps it would choose to linger in Clovelly, a tiny fishing village that is perched on a cliff 130 metres above the Atlantic Ocean on England's rugged south-west coast. Born in medieval times, the village has become a significant tourist attraction which clutches proudly to its past. There are thatched whitewashed cottages with planter boxes and hanging baskets full of the vibrant reds, pinks, and purples of geraniums and fuchsias. Some houses are so close together that it's possible for neighbours to lean out their windows and shake hands. It's a car-free settlement in which donkeys laden with all sorts of cargo, from baggage to shopping goods, laundry to general merchandise, clip along cobblestoned lanes and streets, the longest of which edges down to a windswept harbour and a centuries-old stone breakwater. This steep ramble through the ages is simply known as 'up-a-long' and 'down-a-long'. Like the slow, methodical movement it carries, the road hasn't kept pace with the world beyond the cliff. That role is left to the villagers who come and go. One of those included a man who has lived his dreams by making brave choices at 'the main intersections' of his life.

Bernard Frederick Aust was born in 1949 in a castle. This wasn't a sign of royal blood, but a result of changing times after World War II, when part of Flete Castle in Plymouth was annexed as a maternity unit. Bernie, a round-faced red-haired boy, spent many of his earliest years in transit, stopping just long enough to be placed at the back of a classroom and called a 'squaddie's kid', an unkind term used by teachers when referring to the children of army personnel who rolled from one camp to the next without ever calling anywhere home. Eventually Regimental Sergeant Major Bunny Aust retired from the Royal Artillery and found his family a place to stop rolling when he purchased the Higher Clovelly Post Office and General Store. His two sons delighted in greasing the runners of their father's sledge with lard before loading it up with parcels and packets, and then pushing and pulling the day's load on a door-to-door adventure.

Bernie came to the first main intersection of his life when his father arrived home with an envelope that needed an owner. Initially it had found its way to Clovelly's Postmaster, but neither he nor his daughters were interested in its contents, so it was forwarded to the Higher Clovelly Postmaster and was taken to the Aust home. The eyes of a nine-year-old widened as the letter's contents introduced him to a landscape so vastly different from his own. Although Bernie had already travelled many miles, not even his imagination had dared to roam as far as the place that sprang from the envelope. Within a few days he had licked, sealed and sent his reply on its long journey across the world to Tamworth, New South Wales, Australia, where ten-year-old Richard Hutt was hoping that someone would be curious and eager enough to answer his request for an English penpal.

Over the following years neither boy's interest in his correspondence ever faded. Richard, a musical type with an insatiable wanderlust, wrote of an upbringing in a large country town and district that was filled with dust, heat, flies, summer sunshine, wheat and sheep, kangaroos, droughts, occasional floods, endless flat roads, and weatherboard homes with corrugated-tin roofs: Bernie replied with stories of mud, ice, donkeys, dark freezing winters, herrings, screaming winds, shipwrecks, tight twisting laneways; and houses made of mud, straw, stone and thatch. They sent each other calendars and postcards, photos of themselves and, most importantly, they wrote of their dreams. The more time passed, the more a new dream flourished in Bernie's mind . . . he had to see Australia for himself. It was wrapped in his thoughts alongside his wish that one day he would be a hero who received a medal at Buckingham Palace. Of course, this wasn't an uncommon desire, as a boy can build an entire empire and slay the mightiest dragon without ever moving from his bed.

However, the reality of life in an English village wasn't built on pillows of sleep-dust or falling stars. At fifteen, Bernie confronted another significant intersection when he decided to leave school and begin a six-year apprenticeship as a hand compositor at a small printing office. It was an unhappy time during which he was bullied by the senior workers. Amid sweeping floors and cleaning toilets he did however learn the trade. The hardships of setting words in the workplace were in stark contrast to the pleasure he got from constructing the lines which he sent to his penpal.

After ten years of curling their hands over pages, Bernie and Richard extended them towards each other when Richard went to England for a holiday. The two met at Barnstaple Railway Station near Clovelly. Bernie arrived to find a skinny young man sitting on a bench. He had jet-black hair, and bore a resemblance to the American rocker Buddy Holly. Bernie immediately thought that his mate from afar had 'just walked out of a photograph sent in a letter'. Considering their friendship had already travelled so many thousands of miles, it wasn't surprising that they were quickly at ease with each other. Importantly, Richard's sunny endorsement of his country ensured Bernie's desire to travel 'Down Under' continued to grow. After Richard departed, both men knew their paths would cross again. By the time Bernie had almost secured his work indentures, he had already decided that the most valuable documents he could hold were a passport and immigration papers.

In proof that dreams really can come true, Bernie walked out the migrants' gate at Sydney's Kingsford Smith International Airport on 12 December 1971. He had left behind a heavily frosted day to be met by clear skies and a temperature climbing above 30 degrees, but for a young man who'd built his impressions of a country on twelve years of correspondence, not all was as he'd envisaged. There was 'no dust, no kangaroos, no sheep, and no bloody mate!' Richard was still rushing from the other end of the terminal when Bernie lugged his suitcases and a sudden load of loneliness onto foreign soil. However, after the penpals finally saw each other, any sense of apprehension was comforted by handshakes and smiles, and the prospect of adventures in a far-flung land.

Bernie flew to Tamworth, about 400 kilometres north-west of Sydney, where he lived with the Hutt family and secured a job at a printery. When he wasn't working as a compositor, he was cheerfully interpreting his way through a maze of dropped consonants and nasal drawls that belonged to the down-to-earth, matter-of-fact people of the bush. There were few better places to learn about the locals than in the pubs, which filled every arvo and evening with schooies, snags, counter meals, and a good whiff of bullshit. On one of his early excursions to taste the froth and banter, Bernie was greeted by a barmaid with: 'Whad'll y'ave Strawb?'

'Um, what have you got?' came the polite reply.

'New or Old.'

'I beg your pardon.'

'Aah we've got a bloody pommie 'ere fellas. Jest off the boat are yer?'

Bernie loved it. From the 'No worries' to the 'She'll be rights', from the cracked lips that barely moved, to the corner-of-mouth quips, the man whose profession was to build words was tickled by almost every sentence, even the ones of back-handed affection such as 'yer stupid pommie bastard!' Everything about the Australian way of life intrigued him, not least a barperson's ability to memorise a round of drinks and keep the beer flowing for hours without ever needing to ask the patrons their orders.

This fascination stretched further when the 'bottom fell out of the wool market' in 1972, and Bernie was laid off at the printery. With few other options he joined a council road maintenance gang, where he rubbed sunburnt shoulders with men who knew no way to earn a buck other than by bending their backs and wrapping calluses around tools. They were shearers, cooks, truckies, labourers, builders—blokes just doing their best to make ends meet during hard times. To see a chalk-white pom with his shirt off was a novelty in the stinging heat that prompted more than the occasional joke. In return, Bernie watched with amazement at age-old bush practices that typified a simple life. At lunch breaks it wasn't uncommon to see someone light a fire and cook a steak on a shovel. At the end of most days the gang members loaded themselves onto the back of the work truck to be driven back to town, where they invariably leant against the bar at the Town Talk Hotel. In this environment of sweat and schooners, a fledgling Aussie import learnt much about the values of mateship: work together, play together, and make bloody sure you look after each other. Kindness wasn't a mentioned word but it was never far from the surface, just like the tar that shone in the roads across the Tamworth district.

These roads and a sliding rural economy led to Bernie's next intersection after he was laid off for the second time. Facing bleak prospects in the country, he reluctantly moved to Sydney and contemplated a career with either New South Wales Railways or the Australian Defence Force. However, Richard Hutt's father had another suggestion: 'Why not join the police?'

Until that moment Bernie hadn't any aspiration to be a man in blue, although his brother by then was a member of London's constabulary. But once he considered his options, he thought a life on the beat was more appealing than either trains or a job that would inevitably lead back to a squaddie's existence. So he applied, and to his complete surprise he was accepted. On 11 September 1972 Probationary Constable Aust 'passed out' from the New South Wales Police Training Centre at Redfern. Perhaps fate showed its sense of humour when Bernie was posted to Waverley Station, just minutes away from Clovelly, an eastern suburb devoid of donkeys and sledges but full of bleached manes and surfboards. After he left his homeland he could never have known that his journey would lead him to this point. The only prediction that he packed in his thoughts when he said goodbye was that of his mother's friend, Julie, whose hobby was fortune-telling. She had read Bernie's future three times in her tarot cards, and three times she could only see one clear sign:

'You will meet a dark man at night and it will be very dangerous. You will need to be very careful.'

***

Predictions are open to interpretation, but there was no denying the strength and darkness of Julie's words as Bernie stood staring at a stranger with a shotgun. Thousands of miles away from home a rookie policeman had arrived at the most daunting intersection of his young life. This wasn't a time for bad decisions. He raised his hands above his head: 'Don't shoot. Put the gun down.'

The gunman moved closer and grabbed the front of Bernie's shirt.

'Turn around,' he said in a broad foreign accent.

As Bernie obliged, his collar was yanked tight against his neck, and the gun-barrel rammed into the middle of his back.

'Walk.'

The two men headed towards the shiny, white marble steps that led to the Consulate's front door 20 metres away. Bernie didn't know how many people were inside. After his initial shock subsided, he concentrated on the moment. He had trained for this. When he had taken the oath of office, he passionately believed in his duty to protect and serve the community, even if he had to put his own safety at risk. He was a policeman, and this was his job.

He stopped walking, and resisted the pressure coming from behind.

'Put the gun down.'

'Move!'

Bernie noticed the tip of the gun-barrel out of the corner of his eye. It was pointed at his head and was almost level with his raised right hand. He spun around, knocked the barrel up, and grabbed it. The gun discharged. A shot scorched above his head as he elbowed his combatant in the face. The sound and flash stunned him. He stood still, unable to see or hear properly, yet he was aware that the gunman had tumbled to the ground. Bernie now held the barrel by both hands, and raised it above his shoulders, the stock facing skywards.

'Don't move!' he yelled.

The gunman paid Bernie no heed and sprang to his feet. An instant later Bernie felt a sudden pain in his neck. He'd been struck by something hard. He retaliated, thumping the attacker in the body with the stock of the empty shotgun, and then he dropped the weapon, expecting to see his seemingly defeated rival submit quietly. But the stranger launched again, striking Bernie in the left shoulder. The pain was acute. His neck was throbbing, his shoulder searing. Bernie couldn't lift his left arm. It was only then that he realised he was being attacked with a knife. He tried to reach across his body to his .38 revolver in the holster on his left hip, but he couldn't fight his way through the tangle of twists and thrusts. The two men were so close they could feel each other's sweat, smell each other's breaths. But only one felt something warm and wet pouring down his neck. A strange fizz tingled in his groin and armpits, and flashes of light danced through the darkness in front of his eyes as Bernie fought to stay awake. Waves of dizziness swept in. He panicked: the revolver, the revolver, he had to get to the revolver. With much of his remaining strength he pushed his attacker back with his uninjured right arm and unclipped the flap of his holster. As his fingers wrapped around his weapon he saw the glint of a blade and was then jolted backwards as the knife was driven into the left side of his chest, in line with his pocket. Whether or not adrenaline was shielding him, he didn't feel the sting of an entry. By now he could barely see. In the frightening storm of flashes across his eyes, he pointed the gun in the direction where he thought his attacker was standing. He held his weapon firmly, and pulled back on the trigger. Again and again and again . . .

Bernie fell forward, his head spinning. As he lay on the ground, he could just make out a figure fleeing along the driveway. He sat up and, using his knee for support, took aim and fired. Nothing happened. All six rounds had already been used. The man-to-man fight was over. It was now a battle for each man alone.

Bernie tried calling for assistance on a portable radio, but there was no answer. He crawled down the driveway and inched his way towards the front door of the Consulate. He was unable to stand up. His mouth was dry, his whole body shook, and he gulped in mouthfuls of air. The flashes and fizzes were intensifying. He saw people staring at him through a glass-panelled door from a corridor inside the Consulate.

'Help! Call the police!'

He passed out, and woke lying on his back staring at blocks of apartments that leaned in and swayed over him. He tried to drag himself up the steps to the consulate's front door, but he struggled for a grip on marble that was coated in blood. His blood. He again saw people inside: 'Please help. Police! Ambulance!'

The only movement came from a wounded policeman sliding back down the steps. He began the climb again, and reached the top only to discover that the people he had seen only moments before were gone. He looked at his shirt. It was covered in dark blood from his neck and shoulder wounds. There was no bleeding from the chest; Bernie could thank the police-issue street directory in his pocket that had saved him.

But the road he was trapped on was grim. As he slumped against the steps, he saw only one clear direction in which he could go. He thought he was about to die. Yet still he held on to his gun, believing he had to protect the Consulate. As far as Bernie was concerned, he would perform his duty to the very end.

He heard footsteps rushing towards him. He raised his weapon, but lowered it when he recognised the voice of a colleague. The officer bent over and checked the wounds before running off again.

'Come back! Please don't leave me,' Bernie yelled as the footsteps faded away.

Other footsteps hurried closer. Bernie again lifted his revolver before he heard an unfamiliar voice: 'Don't shoot! Can I help you? Can I help you?'

It was a civilian who'd heard the gunfire from his nearby home. From his balcony he'd seen Bernie struggling. He arrived and, after a brief inspection, he tried to stop the blood flowing from the most serious injury; Bernie felt fingers dig into his neck. But then the pressure eased, and was soon gone altogether as the man hurried away in search of a doctor.

'Please don't leave me!'

Bernie's body tingled all over. He coughed as he fought for every breath. He was sleepy, weak, and very, very lonely. What a terrible way to die, so far away from his family, from the people he loved most, from his home; 'no one deserved to die alone'. All that he could do now was pray to God and wait.

At that moment he was calmed by an overwhelming sense of peace. All panic had gone; he had accepted his fate . . .

Commotion!

Bernie's peace was broken by the arrival of a doctor and a police officer. Others were soon swarming at the scene.

Questions poured out. 'What happened?' 'What did the attacker look like?' 'Where did he go?' 'How did it start?' . . .

Among the few words Bernie registered were the ones that confirmed his fate as he was rushed away on a stretcher: 'Go with him and get a dying declaration.'

By the time Bernie arrived in the Emergency Ward at St Vincent's Hospital, every second was frantic. Blood continued to flow from his sliced jugular vein. As a doctor clamped the wound, Bernie writhed in agony, forcing a number of nurses to throw themselves across his body. In the minutes that followed he slipped in and out of consciousness, and somewhere in this drifting mist, danger passed him by. He was going to survive.

The next morning his attacker was found dead near a block of apartments not far from where he first attacked Bernie. A post-mortem revealed that 27-year-old Odicho Sleiman Younan, Lebanese by birth, had been shot in the chest and had probably bled to death within 30 minutes of sustaining the injury. The lethal bullet had been fired by Probationary Constable Bernie Aust. Police investigators quickly learnt that Younan had a criminal record in both Victoria and New South Wales for offences including: Wilful and Unlawful Damage, Malicious Injury and Vagrancy. Deemed as 'dangerous and mentally unbalanced', he was wanted by Victoria Police in relation to an attempted murder in which he allegedly used a knife. He had recently sought assistance from the Lebanese Consulate in a bid to return to his homeland, and had worried officials with his agitated behaviour.

A coronial inquiry cleared Bernie of any wrongdoing. In a twist of coincidence, the findings were announced on another 'Black Friday'—13 April, 1973: 'It is clear that Constable Aust would have merely apprehended the deceased if he had been able to, but that proved impossible. The constable has shown excellence in the performance of his duty and displayed a great deal of courage in the face of the attack.'

* * *

In the days after the event Bernie was front-page news, but no headlines or high praise could soften the impact of a specialist's prognosis: 'I don't think you'll ever be able to talk properly again.'

A vocal cord had been mutilated beyond repair. There was also substantial nerve damage, and Bernie's left jugular vein would never work properly again. The rehabilitation was going to be slow, but just four days after the incident Bernie was prematurely discharged from St Vincent's after the hospital received some anonymous death threats directed at the recuperating policeman. Bernie's flatmate received similar calls, and because of them moved out. Amid fears for their young officer's safety, police officials allowed Bernie to convalesce in Tamworth with the Hutt family.

So, it was back to the bush for a nearly mute man who soon found that the loudest statements he heard were in his own thoughts. He slipped into a series of lows as his conscience was tormented with guilt. Had he done the right thing? Did he really need to shoot the man? What did it really mean to kill someone?

It didn't help that nearly everyone he met recognised him and either wanted to congratulate him or discuss the incident. Bernie was forced to re-live the night over and over and over again. However, in this horrible loop of 'a tape that wouldn't stop running', came long-playing kindness. Trish Smith was a friend of the Hutts. She was a kindergarten teacher whose earthy nature reflected her upbringing on a farm. She'd met Bernie only once before, at a dinner party, but after she heard of his misfortune she visited him, bringing the treasures of a warm smile and compassion. She started taking Bernie for drives around the district. She showed him the sights, introduced him to new friends, told him stories about the ins-and-outs, and in-betweens of Tamworth. Days passed, miles clicked over, weeks passed, wheat crops were stripped, sheep were shorn, dust rose and settled... Against this backdrop of bush life, the seeds of caring grew into a romance. In the two or so months that Bernie stayed in Tamworth, he wasn't once asked by Trish about the incident at the Consulate. For two young adults falling in love, the future was all that mattered. Not even the challenges of a long-distance relationship troubled them when Bernie returned to restricted duties at Waverley. On days off he regularly flew to Tamworth, or else Trish visited him.

Bernie's return to duty was met by some hardship and much humour. He was only able to communicate in soft breathy whispers, or by writing on a notepad. Officers who didn't know his circumstances occasionally chastised him for his seemingly slow and meek responses but once they were told the truth, ignorance made way for admiration and sympathy. Others, most notably detectives, took great delight in using their impaired colleague as a runner, asking him to 'go tell so and so that he's needed on the phone'. This generally meant a dash upstairs, followed by an almost silent, breathless message accompanied with hand signals, while good-natured chuckles rumbled from the floor below. One day, about three months after the Consulate incident, Bernie was in a room full of detectives when his words suddenly roared to life: 'Sergeant, there is a PHONE CALL FOR YOU.'

Fantastic! Wonderful! You bloody ripper! Bernie's voice had unexpectedly returned.

At his first opportunity he flew to Tamworth, and was greeted at the airport by Trish.

'Listen to this I can talk again!'

'That's great, but *shhhhh*, there's no need to shout. Everyone can hear you.'

Bernie had become so used to needing all his breath to force a sound that it took him some weeks to adapt to speaking normally again. And when he did, he confidently asked the most important question of his life:

'Trish, will you marry me?'

Bernie Aust and Trish Smith were married on 1 September 1973: the first day of spring. Soon enough, they bought a simple weatherboard house, and began raising a family: their children Brian, Josephine and Dominique were born just three years apart.

Five months after he was married, on 2 February 1974, the Queen Mother presented Bernie with the George Medal, one of the highest honours for gallantry awarded in the British Commonwealth. To this day, only six New South Wales police officers have ever received the award. (Five were serving as police officers, the sixth was a soldier at the time.) Although modesty prevented him from thinking he was a hero, Bernie had indeed fulfilled another childhood dream—the ceremony took place in Buckingham Palace.

For someone who had thought he was dying less than two years earlier, life was truly worth embracing. This included the police work that took Bernie on a 'wonderful adventure'. In 1974 he became a detective who spent four years of 'solid, hard graft learning' before he reached another of his life's intersections when he accepted an invitation to join the hand-picked Crime Intelligence Unit (CIU). This was at a time when organised crime and police corruption was rife. In one of his roles with the CIU, Bernie was a surveillance person, travelling all over Australia following some of the most legendary criminals in this country's history. His duties included driving a VW Kombi, parking it at various locations, and climbing into the back where he took photos behind the cover of curtains. Such clandestine operations gave Bernie a 'fascinating view of the underworld' at a time when surveillance work of this type was still in its infancy. The click of the shutter introduced him to such notorious types as George Freeman; Darcy Dugan; the Trimboli family; and Lennie McPherson, the so-called 'Mr Big' of Australian crime, whose network stretched from political corridors and police stations all the way to the Mafia and the CIA.

Bernie also worked on secondment with the National Crime Authority, and on his return to the New South Wales Police Service he established the Surreptitious Entry Unit, a highly specialised team whose duties included picking locks, defeating alarms, and installing listening devices in a wide range of premises. It was thrilling work.

Eventually Bernie moved into supervisory roles, and today he is in management as Chief of Staff to the New South Wales Commissioner of Police, Ken Moroney. Despite all the years that have passed, he still lives with the memories and the results of Black Friday, October 1972. To kill, and almost be killed in such a manner was always going to leave scars. It took him seven years before the 'tape stopped running', but to this day he still has flashbacks, most notably every year when he undergoes compulsory firearm assessment at a shooting range. Bernie also has a number of physical problems that he has been forced to accept and adapt to. Such is the damage to some of his neck nerves that he doesn't experience hunger, nor the sense of satisfaction that most of us have after eating. If he tries to sleep on his left side, he suffers severe headaches caused by his blocked jugular vein. He has never regained full movement of his jaw, and he perspires on only one side of his face. And yet that face has an endearing smile and cheerfulness that reflects a genuine contentment.

Bernie Aust has been a police officer for more than 30 years, and has now begun to think about his next main intersection—his retirement. He has no doubt about what direction he and Trish will head. It will be north-west, where they plan to buy a piece of dirt in the Tamworth district and run a few cattle. It's the 'real Australia' Bernie yearns for. When he begins this new adventure, he will be back in the place where his Australian life began and his close friend, Richard Hutt, still lives in the area. The two former penpals have now been mates for nearly fifty years.

'I've come to many big intersections in my life, and the choices I've made have taken me on an incredible adventure,' says Bernie. 'There was the letter, being bullied in a trade, emigration, the bush, the police, the incident, the Crime Intelligence Unit. All those things set me on this course. And along the way I met the people who have made my life what it is. There's Richard, we will always be the best of mates; there's all the characters of the bush; the wonderful people I've worked with; the scoundrels and criminals; my children; and Trish, who is the most important person who has ever come into my life and is the best thing that has ever happened to me. She took the time and trouble to find me. She is just an incredible lady. And to think it all started with a letter when I was nine years old. That set me on the way. My life in this country has just been one huge adventure. Australia is the best country in the world for adventure. I have been very lucky.'

And there, for the moment, ends the story of Bernie Aust.

KEVIN 'BILLY' BOYLE

'Down the Street'

'Billy is so passionate about the foundation that he can almost be a pest.'

Command District Officer Greg Crossman, South Australian Metropolitan Fire Service

'What do you mean almost?'

Kevin 'Billy' Boyle, Senior Firefighter, South Australian Metropolitan Fire Service

A few years back, this bloke was riding his motorbike to work in Adelaide, 60 kilometres away from his home town, Springton. It was his first day back on the job after a holiday. It was August. And it was freezing. This bloke only went about twenty k's before the chill forced him to pull over to the side of the road and rummage through his bag for anything that would keep him warm. He reckoned his 'hands and legs weren't too bad, but my face was bloody burning!' He soon found something that with a bit of innovation could shield his skin without obstructing his view through his full-face helmet. He continued on his way, zinging along until he pulled up at a set of traffic lights not far from his workplace. Merrily minding his own business, he turned to his right, and was surprised to see the man in the car next to him staring at him in amazement. It was certainly a justifiable reaction because it's not every day you see a person wearing a pair of underpants on his head.

'But they worked!' says this bloke, giving no doubt that if the need arose he would don the jocks again. 'I pulled the trunk hole over my face and looked out through a leg hole. No problems.'

This bloke is Kevin Boyle, a man you're unlikely to forget once you've met him. He has carried the nickname 'Billy' ever since he was at primary school, and now 40 years later, few people know him by any other name. Billy still looks fit enough to lace up his boots and play a game of his beloved soccer. He is of medium height and athletic build, although he has a slightly stiff gait resulting from an old knee injury that his favourite sport farewelled him with. He has dark, short-cropped hair, a chiselled face with sharp hazel-green eyes, and he occasionally speaks out of the side of his mouth as though he's giving you a quiet tip for the fourth at Flemington. However, he isn't a gambler; when you are as charitable as Billy you know there are too many better ways to use money.

He is a member of the Australian Professional Firefighters Foundation (APFF), an Adelaide-based organisation that's operated solely by firefighters and their families. Since forming in 1998, it has raised hundreds of thousands of dollars for countless causes, and in recent times, no one has worked harder, longer, and more passionately than the eternally energetic Billy Boyle.

His appreciation of those in need began when he was growing up in Elizabeth, in Adelaide's north. It was purposely built in the 1950s by the South Australian Housing Trust to accommodate the influx of post-World War II immigrants from Great Britain and Europe. Billy's parents were 'ten pound tourists', arriving in Australia in 1962. His old man, Harry, a Scot, used to drive trucks across the length and breadth of Britain. On one trip to England during the war he met his future wife, Doris. They married in 1950, and by the time they arrived Down Under, they had one daughter and three sons. Billy was the second-youngest.

While Doris looked after her children (she would later have a second daughter), Harry secured a job driving trucks at the General Motors Holden's plant, the enormous hub of Elizabeth. After first living with relatives the Boyles moved into a typical trust home, a Besser brick duplex with three bedrooms, a shed at the back, and a tidy front lawn with a two-strip concrete driveway and a row of oleanders that acted as a fence between properties. However, it was a move that was almost scarred by tragedy. Billy recalls:

'53 Heytesbury Road, Elizabeth West. It was a new house. My old man and uncle went to inspect the place before deciding to move in. They parked their little Austin Cambridge in the driveway and told us kids to stay in the car. But, you know, being kids we weren't meant to do what we were told, so we got out and started playing in the front yard. It was lucky we did 'cos a drunk driver came along and wiped the old Cambridge out. Could have killed us all.'

This was among Billy's first vague recollections of living in the town named after Queen Elizabeth II. But far from enjoying a regal existence, the Boyles rarely had any money to spare. The simple equation of the household was: money in equals money out. The children weren't lavished with toys or trips, and they were dressed in hand-me-downs and second-hand offerings that Doris bought from a St Vincent de Paul's shop in which she worked as a part-time volunteer. Her youngest boy would soon enough inherit her devotion to others less fortunate. Harry and Doris sacrificed much for their children, and it wasn't until years later when working as a third-year apprentice at 'Holden's', that Billy understood how tough it must have been for his parents. By then, as an eighteen-year-old, he was earning more than his father.

Prior to this, Billy had weathered his adolescence, a period of schoolwork, sexual flirtations, kicking soccer balls, and hanging out as a 'non-violent hood'.

'I suppose I was like any young kid looking for things to do in an area that didn't have many distractions. When I was about fifteen I started hanging out in a gang near the local shops. We weren't into house break-ins, or anything like that. We'd mostly sit on the corner and pretend we were tough. It was just something to do. When I was seventeen I got pinged for under-age drinking, and had to go to court. The old man had to take a day off work, which didn't make him happy, but when he stood in the dock he went into this spiel about what a great kid I was. I still got fined though.'

Harry played a strong hand in guiding Billy at this time. After his son left school in Year 11, he was adamant that the logical path of progression led to 'Holden's', as it was popularly called. And that is how Billy, somewhat reluctantly began his four-year apprenticeship as a maintenance fitter. But his aversion was soon swept away by enthusiasm when he discovered that in a workplace dominated by British migrants, the favourite topic of discussion was soccer. By this stage, Billy was a star on the rise in the local ranks. He was eighteen when he made his first division debut as a striker for Enfield Victoria in the State league. He would later captain Salisbury United, and be chosen to play a trial against an Australian World Cup squad before cruciate ligament damage cut short his career when he was just 27.

Only a few short years after hanging round on a street corner, Billy suddenly found his life was full of activity. Work would always be work, but outside Holden's gates, he clicked into gear. He played hard on the field, and even harder off it, earning a reputation as a social bloke who would talk to anyone, and if there was a drink to go with it, well, it was best to keep them coming all night. It was during this time that he 'hooked up' with Linda van Meerendonk, a girl whose parents were Dutch immigrants.

'Coming from the same neighbourhood, we'd seen each other round for years,' says Linda. 'When I was seventeen I knew I wanted to marry him. We talked a lot. Billy loved to talk. Talk, talk, talk, talk, talk, talk.'

They were engaged in 1978, and married two years later. Billy was 22, Linda 20. After living their first two years together in a trust flat, they had saved enough to build their own home in the area, but it was a difficult time. While Linda's employment as a bank clerk was stable, Billy was worried, as rumours spread that Holden's was going to close, or at the very least reduce its workforce. One evening when he was coming home from soccer training Billy watched some firefighters battle flames at Smithfield Railway Station, just a few minutes away from Elizabeth. Standing there in admiration of the men at work, he started thinking: 'You know the firies wouldn't be a bad job. I wonder how I can get in.' The very next night he remembers good fortune shook his hand when he met firefighter Terry 'Butchy' Butler at a bucks party.

'We got talking about the job, and the more he told me the keener I got, but I had to wait for the next round of applications to be advertised. I asked Terry to let me know when they'd come up. A couple of days later he rings me and says: "Get the paper!"'

Billy thought he could satisfy most of the criteria, but there was one noticeable obstacle . . . he didn't have a truck licence. As a boy he'd never had any yearning to do the same job as his father, but in the hurried weeks before applications closed, he was kept busy crunching gears at a driving school.

It was a 'stinking hot' afternoon when he arrived at the local Motor Registry Office for his practical test, but bad news greeted him: 'I'm sorry,' said the booking clerk, 'your examiner has cancelled, so we'll have to postpone it.' On any other day it wouldn't have mattered to Billy, but this time urgency was needed because the cut-off time for lodging applications was just three hours away.

'Shit!'

Again good fortune offered its hand when, while Billy waited, the clerk took a phone call from another learner driver who had to cancel his test. Billy stepped into the vacant spot. During the test Billy tried to make conversation with the examiner:

'I'm really glad I'm going for this licence today.'

'Why's that?'

'I'm trying to get into the fire brigade.'

'Oh, my son's going for one of those jobs too!'

He passed. The newest qualified truck driver in Adelaide hurried into the city and lodged his application at Brookway Park, the training college of the South Australian Metropolitan Fire Brigade. It was 4.25 pm, only five minutes before the cut-off.

'Looks like you'll be the last one,' said the woman behind the counter. 'Good luck!'

On 18 July 1983, Kevin 'Billy' Boyle became a recruit firefighter. After fourteen weeks in 'Drill Squad' he drove home with his uniform on, feeling 'bloody proud to be a firie'. He was posted to Elizabeth, where he embarked on his journey 'down the street' the term used in the brigade for being on the job. He continues working at Elizabeth to this day. The diverse memories he has collected over the years only truly become apparent when he gives you a running commentary while driving through the area.

'See that vacant block over there? It used to have a house on it. Some kids set up an LPG cylinder in a wardrobe. They turned it on, lit it and hoped it would explode. The house got burnt down and bulldozed. When that happens we call the vacant block a car park . . .

'And that house over there, used to be run by drug dealers. You get a lot of hydroponics round here. You get a call-out to a fire, and there'll be power leads running everywhere all over the house . . .

'This house here had two murders in it only a few years apart. Two different families and different circumstances . . .

'There was a brothel down here . . .

'We dragged an old hobo out of a fire there. We tried to revive him but he was already dead. A whole lot of kids were just standing around watching us, saying: 'What's wrong with him, mate? . . .

'And we get a lot of trouble in this street. There's a load of kids live in that house. They have bonfires, get pissed, get in their cars and do burn-outs, people see the smoke and call us. There's always something happening here . . .

'Now this house is a good one. I wasn't on duty the day it happened, but the crew get a call out to a rescue and find this bloke totally butt-naked chained up with zip ties and chains. And you know what the final part of his fantasy is? To be rescued by a fireman! The boys came back to the station, and reckon they were scrubbing their hands for the rest of the shift . . .

'And down there, a transvestite used to live down there. And you ask me why I work here? Well it's really interesting!'

Elizabeth is a tough area where unemployment, drugs, alcoholism and teenage pregnancies all pose problems. One day when Billy visited a school to promote the Fire Brigade he asked a boy what he wanted to be when he grew up.

'I want to be on the dole like me father and grandfather. They get money off the government for doin' nothin'.'

By that time Billy was a father himself and listening to this child made him realise that although he had a growing desire to help others, he was a staunch believer in the age-old adage that charity began at home. When he wasn't working he was invariably learning the ways of parenthood with Linda and their son, Damian.

After spending some time in Springton, he and his family moved to Lyndoch, a town 50 kilometres north of Adelaide at the southern end of the Barossa Valley wine-growing region.

'Country life was always something Linda and I wanted to do. We wanted somewhere full of space for our kids to grow up. Somewhere we could grow up together as a family,' says Billy.

By 1992, the family had grown to five, with the births of Meagan, and the youngest, Chelsea who arrived three months before Billy's father died from Alzheimer's disease. When Doris Boyle found life as a widow difficult, Billy and Linda built a 'granny flat' at home, and together with their children, looked after her during the final four years of her life. At her funeral, the renowned bond between firefighters was underlined when a group of Billy's workmates arrived at the church in a station truck to attend the service in full uniform. None of them had ever met Doris before. Billy was overwhelmed by their kindness, it was an act that reinforced his belief that 'family and friends should be there for one another. You have to support one another through the good and the bad.'

No one could argue that Billy practised this. As the children grew older, he and Linda were the ones regularly seen driving a car-load of kids to soccer matches, turning the sausages at school barbecues, cutting sandwiches in the tuckshop, building sheds, and generally offering a helping hand or strong shoulder to any community project. In this spirited town there was another firefighter, Paul Rayner, who was just as generous. He and Billy reflected the motto of many firies: 'get in, give a hand, get the job done'.

Billy also helped re-establish Little Athletics in Lyndoch. He and a few other parents undertook official coaching courses, and then passed their knowledge and, most importantly, their enthusiasm, on to as many as 80 kids who ran, jumped and threw until darkness chased them from the track each Friday afternoon. At the same venue, the Lyndoch Primary School Oval, Billy turned running lanes into golf fairways for students he worked with as part of the school's Learning Assistance Program. This involved adults volunteering to be mentors for children. Billy felt it was his 'responsibility to care, 'cos it's important to help others. Experience helps too. I suppose I'm a bloke who's done a bit of living.'

His strong sense of obligation still drives him at a furious pace at home and in the Lyndoch community. (At the time of writing he is coach of the Barossa United Under 15s girls soccer team for which Chelsea plays, and he does countless odd jobs, ranging from mowing lawns and labouring, to cutting the toenails of 'Dear Mrs C', a distinguished elderly woman who once stood alone in an angry crowd at the Adelaide Oval and clapped infamous English cricket captain Douglas Jardine, the mastermind of Bodyline, because she thought it was 'beastly to boo like everyone else'.

However, it's as a member of the APFF that Billy is a true whirlwind of activity. After joining in 1999, he did little apart from donate two dollars per fortnight until he attended a house fire just before Christmas the following year. The house was destroyed after it had gone 'scone hot'. Afterwards, Billy noticed a young girl who sat weeping in the gutter. She was one of two foster children being cared for by a single mother who also had four children of her own.

'You can try not to get involved in these things, but sometimes your feelings take over. This poor girl, she'd probably already had a lot of trouble in her life, was about the same age as Chelsea. People get bad breaks they don't deserve, and this little girl was one of them. No one deserves that at Christmas.'

When alerted to the incident, the APFF allocated $600 to buy presents and provisions for the family, but Billy still wanted to do more. Armed with a newspaper article about the fire, and with daughter Meagan, then twelve years old, he approached shopkeepers and department store managers for donations. As word spread, the public dipped into its pocket, leaving offerings at the Elizabeth station. Eventually $3500 worth of food and gifts were wrapped by the firefighters and delivered to the family only four days before Christmas. For Billy, 'seeing the joy in those kids' faces brought a tear to my eye'.

It also opened his eyes to how much else could be done. From that moment Billy became heavily involved in the APFF, and was soon among the Foundation's driving forces; his tireless efforts to raise funds earned him the nickname 'The Stalker', as he wasn't afraid to ask anything of anyone in the name of charity.

This stalking and enthusiasm continues to grow; if Billy knows you, there will come a time when he inevitably asks you to support a cause. Maybe it's to help Amber, a girl with a life-threatening brain tumour who needs to fly to the United States for special treatment; or perhaps it's to buy blackboards and toys for children with burns injuries in the Newland Ward at the Adelaide Women's and Children's Hospital; or it could be to purchase consumables for the Skin Engineering Laboratory at the Royal Adelaide Hospital; or provide books to the Davoren Park School Library which has lost 14,000 items in a fire that was deliberately lit.

And then there's the special case of 'Old Ray', the 81-year-old who is terminally ill with cancer. Two years ago he was found highly distressed in his yard while his unit burned behind him. Although the fire was brought under control, Ray, who lives alone, lost many of his possessions. With support from the Anglican Church charity, Anglicare, Billy went corporate door-knocking. Ray is now living his final years in comfort thanks to new clothes, a new bed (complete with quilt and linen), a new radio/CD player, a new heater, and a smart new electronic recliner chair in which he sips his brandy when Billy comes to share a drink with him every few weeks. Billy shakes his head when acknowledging: 'It's no hardship for me to go and see him; I really enjoy his company, but the problem is there are too many Rays out there.'

If Ray's case reveals Billy's heart, 'Camp Smokey' shows his soul. It is an annual four-day camp for children between seven and sixteen years old who've been treated at the Women's and Children's Hospital for all types of burns. It was established in 1991 by a group of nurses who were committed to improving the lives of their patients beyond the treatment they received in their wards. Nowadays, the nurses and firefighters come together for the camp. As many as 40 children attend, participating in various sports and outdoor activities such as hiking, rope climbing, swimming, golf, and even treasure hunts. The camp also plays a crucial psychological role in helping the burns victims overcome any fears of fire, and it's here that there is no shame or embarrassment in bearing scars.

Billy began attending the camps as an adult helper in 2003. Again, he was affected by what he saw, acknowledging 'the enormous courage of the human spirit'. It's not surprising that Billy has been involved in every 'Camp Smokey' since then. When he's not piggybacking a giggling child or pushing one on a swing, he's stalking for memorabilia items to be auctioned at the APFF's annual charity ball, which raises money for the camp.

'My only worry is that he will tire himself out,' says Linda, who Billy readily refers to as his CEO (Chief Executive Officer). 'I have to force him to take time out, just sit on the couch with the family, and try to relax. Or I have to take him on a break for a weekend so he can really get away. Do you know he loves fishing?'

The image of Billy with a rod in hand sitting peacefully on the end of a jetty at sunset is splashed with incongruousness. Could it really be the same man who has been known to own as many as three mobile phones, and have the ability to hold separate conversations on each at the same time? He remembers one occasion when he answered a call on his APFF phone, but found no one at the other end. It was only afterwards that he realised he'd bumped his personal mobile, and had rung himself.

In truth, Billy Boyle invariably answers to anyone but himself. As Linda suggests, 'Billy wants to help everyone else all the time. He'll have about six million thoughts going through his head, but none of them are about him. He just has this really strong need to look after others. What else can I say about him, other than I love him.'

The final word in this story undoubtedly belongs to the indefatigable bloke who knows how warm a pair of jocks on the head can be. That freezing day all those years ago in the Adelaide Hills was one of the few times that this bloke stopped long enough to help himself. Looking back on all he has done, he reckons, 'I'm not special. I'm just being myself. I'm your typical Australian who likes to have a good time. I like to be happy, I suppose, and I like others to be happy too. I am lucky because I have a beautiful wife who always supports me. The CEO gives me stability. I'm not the perfect father, but I've tried. I know I should hug and kiss my kids more, but I have tried to be there for them. It frustrates me the parents who never want to see their kids run 100 metres or don't watch the school plays or band nights. We've got to be there for each other whenever we can. Not just family, but everyone.

'The South Australian Metropolitan Fire Service has been really good to me. I've been hit by how precious life is. I've never counted how many deaths I've seen, but it's been quite a lot. I go to motor vehicle accidents and ask questions to myself: "Why was this young driver pissed doing 100 k's?" It hits home how it can all be over before you know it. But there's the bloody good side of this job too, you know. I love it. I'd never have a sick day for the sake of having a sick day, and I never say to my wife I don't want to go to work. I like the guys I work with. There's a mateship, a lot of different bonds happen at the station. If we do go 'down the street' and cut a person out of a car and help save a life, or if we save someone's property or stop someone's house from being flooded, well that's a bit of magic. It's nice to make people happy. A lot of people say they don't know we're there until they need us, but when they do, we handle the jobs with dignity. It's a pride job.

'The Foundation (APFF) is an extension of that. Just to see people smiling and laughing when we've done something for them, it makes me happy too. That's what it's all about.

'I know I'll keep working for a while yet, but I wouldn't want to be 62 and getting out of the truck in a walking frame. When I do retire I know that I just couldn't sit at home. The more you do, the more you want to do. If you know what I mean.'

We do, Billy.

NB: With special thanks to the boys on Elizabeth Station C-Shift: John Harris, Bobby Nairn, Chris Nottle, Bob Fraser, Collin Ween, Cameron McKenzie, and Leon Shepley* (*normally D-shift).

THE MFB

'38 C-Platoon'

'When I thought of the MFB before I got in I'd look at the guys and think "wow!", but now that I am in, I've realised they are just everyday people doing a job that is a little different.'

Greg Plier, Qualified Firefighter,
Metropolitan Fire and Emergency Services Board

The life of the firefighter isn't about a hero striding out of a burning building with a baby under each arm and a coughing parent slung over a shoulder. That is Hollywood. In reality, most firefighters will never rescue a person from an inferno, but their roles remain greater and more varied than those portrayed on the silver screen.

The South Melbourne Fire Station provides a classic cross-section of modern-day firefighters. It is one of 47 stations operated by the Metropolitan Fire and Emergency Services Board (MFB). Known as Station 38, South Melbourne primarily services the CBD and inner suburbs, which have a mix of high-rise buildings, private dwellings, industrial and commercial plants, hazardous material depots, ports, and extensive transport networks. This diversity ensures a wide array of responsibilities. The majority of South Melbourne's work is 'turning out' to alarm calls, many of which originate in high rises. Upon investigation, these calls are generally found to be false alarms, where the triggers may be as simple as faulty wiring, a leaking sprinkler, or the result of a person swinging from and breaking a water pipe. These incidents are relatively mundane and certainly don't lift the firefighter's heart rate as the 'big burns' do. All the members of Station 38 have their own stories of memorable blazes they've fought, ranging from incidents at suburban weatherboard cottages to Victoria's largest chemical storage facility at Coode Island. And yet, their duties don't just involve defeating flames. On any given day, a firefighter may be called out to pull a trapped person from a motor vehicle accident; clean a chemical spill; break into a house for an owner who has locked his keys inside; cut down a person who has hanged himself; and yes, even rescue a cat whose sense of adventure has taken it to the very tips of a eucalypt.

The roles at Station 38 extend even further to include the operation of special equipment that monitors the air for hazardous materials. Not only is this used at fire and accident scenes, but considering we are now living in the age of terrorism, its importance spreads far beyond everyday occurrences. Furthermore, those at Station 38, as with all MFB platoons, are part of the 'First Responder' Program, which means firefighters are dispatched to cardiac arrest cases (a person stops breathing and has no pulse) at the same time as the closest emergency ambulance is. When the firefighters arrive first, they start work on the patient immediately. All this can lead to a varied work day, but not necessarily one that has the mass appeal of a Ron Howard blockbuster.

Station 38 is the second-biggest station in the MFB. It has two primary appliances (A and B pumper trucks), and three secondary appliances (a breathing apparatus vehicle; a support vehicle, and a truck with a ladder platform that can reach a height of 34.5 metres). There are a minimum of thirteen firefighters on any one shift, each of which is filled by one of four platoons: A, B, C or D. Each platoon will work two successive days (8 am to 6 pm), then two successive nights (6 pm to 8 am), before having four days off. Each firefighter is assigned a specific duty for every shift. Ranks are determined by time in the job, and additional training qualifications.

The process of 'turning out' usually begins when an emergency call goes to the MFB's Communications Centre, where details are loaded into a computer system. From here, the computer automatically assigns the station and appliances. After a dispatch button is pressed at Communications, the order is sent to a computer in the appropriate station's watch-room. This in turn goes into a unit that switches on all the lights in the station and sounds a two-tone alert (gone are the days of bells at many stations). At this point, the station's front doors go up, and an announcement allocating appliances is made from Communications over the station's speakers. This is repeated twice. The appropriate crew members then dress, and prepare their equipment. One person, generally the Station Officer, will pick up a piece of computer print-out from the watch-room which has details of the incident. Once the crew is in the appliance, the Station Officer uses a radio to inform Communications that the appliance is on its way. The normal response time from the moment the 'tones go off' to the whole crew being in the appliance ready to go is less than 90 seconds.

In a standard fire situation, at least one pumper truck turns out to every incident. There are four people in a truck: the

Station Officer, who travels in the front passenger seat and, once at the scene, is responsible for all decision-making; the driver, who is also in charge of pump operations and acts as a conduit between Communications and the Station Officer; and two firefighters who, after riding in the back section of the driver's cabin, perform jobs determined by the Station Officer, such as rolling out the hose lines and spraying water on the fire. This is a very simple breakdown . . . the rest is best told by the firefighters themselves.

While writing this book I was invited to spend a weekend on shift with South Melbourne's 38 C-Platoon. They are a mix of married, divorced and single men and women ranging in age from the mid-twenties to the mid-fifties. Many have come from other jobs, significantly trades, while others have journeyed to the front line via classrooms, gyms, helicopters, and even the sets of the long-running television series 'Neighbours'. They are a lively bunch who, when possible, follow the MFB's tradition of enjoying Sunday roasts in the mess. It's here that banter and bluster flow more readily than the gravy.

The following interviews are snapshots of their working lives. Some of their accounts are graphic. As author I make no apology for this because their matter-of-fact descriptions underline to us how easily we can take their roles for granted.

ANDREW O'CONNELL

SENIOR STATION OFFICER
36 YEARS OF AGE

I was only twenty when I joined the MFB. I've been in it for seventeen years, and I'm staying here; lock it in. It's like a club. Short of flying a space shuttle there's no better job.

There is a great bond between firefighters. If I fall over in a house fire, someone has to come and get me, I hope! There is a strong trust that you don't question. Because you work so closely together, your life becomes part of everyone else's as well. There aren't many other jobs where you and your workmates see and share the full range of human emotions. We see the best and the worst. It brings us together.

The culture at the station is really interesting. You have to have a pretty thick hide to fit in. When new guys come in, we always test them out to see if they have any weaknesses. Put it this way, if we call you 'boxhead' and you don't like it, you're going to be 'boxhead' for the next 30 years. Practical jokes are big, and so is black humour. People outside wouldn't understand but black humour is a very important way for many firefighters to deal with what they come across. It's a release.

You can tell when stations are due for a fire because people start to get a bit antsy. But then you turn out to a big incident and when you come back everyone is full of chat. You hear comments like, 'Hey did you see me on the roof with the hose. How good was that?' They can talk about it for weeks and re-live it. We also talk about the bad ones too, which is something we need to encourage because keeping those to yourself can be one of the worst things you can do. That's where some of the more experienced guys are good because they are happy to sit at the mess table and talk about it. We also have a thorough and highly professional Critical Incident Stress Program that allows any of us to seek further help should we need it.

One of my biggest memories is of a house fire, I don't know how many years ago now. When we arrived we were told there was a baby in the bathroom, and the mother had gone back in to get her. We had a quick look around the outside of the house for louvre windows because that's often where a bathroom or toilet may be, but it was discovered the toilet was inside. We went into the house [Andy and partner] and heard the mother struggling in a corner, so we found her and dragged her out. The strange thing about her was that she had a prosthetic leg, which I obviously didn't know about at the time. As I was dragging her out the front door I noticed her feet were pointing in opposite directions, and I thought: 'How the hell have I done that to her?' Despite what you see in the movies, there's no fireman's lift in these situations. I just grab your arm and drag you out because I don't have a great deal of time. No one has ever died from a bump or a bruise. So we got her out, and she was unconscious. We gave her to a couple of firefighters who started work on her with oxygen and resuscitation gear.

Then we've gone back in to look for the baby. We found her on the bathroom floor, and unfortunately she didn't look good. She'd vomited. We took her outside, and the ambos had just got there. The first guy had knelt down by the mother to try to give her some air, but as soon as I came around the corner you could literally see the colour drain from his face as he saw this little baby in my arms. They worked on her for ages, and in that time an enormous storm came through, but I had no idea it was raining until the guys put a tarpaulin over our heads. It was the only time that ever happened, I was so 'in the zone' that I didn't know what was really happening around me.

The mother survived but the baby didn't make it. She was wearing exactly the same singlet that we had for one of my little girls. When I got home I asked my wife to throw it out; it was something I didn't want to be reminded of too many times.

Firefighting can be frightening. You go to some of the larger chemical fires where 44-gallon drums explode and scare the living daylights out of you. You're doing your job then all of a sudden... *BANG*! And one of the drums will just take off.

Coode Island is the biggest fire I have been to. It was a big toxic chemical fire that burned for two days. I can remember seagulls would fly along into the smoke and fall out of the sky. Dead! I've never seen that before.

Being a firefighter is dynamic and full of challenges. I love coming to work.

LOUISE CANNON

STATION OFFICER
44 YEARS OLD

I was originally a Phys Ed teacher, but hated it and needed to look for something else. I had no idea what a firefighter did, and I probably joined for all the wrong reasons. You could get four days off a week and the holidays were good.

I was the fourth woman in the service. I was 28, and very naive. I've now been in for sixteen years, nine at South Melbourne.

When I first came out as a recruit I was sent to another city station. There were 50 other guys and I was the only female on all the four shifts. It was different from the way it is now. There were guys there who were confirmed woman-haters, who said they'd never get in the truck with a woman, and there were wives who didn't want a woman at the station.

The introduction was tough. A couple of weeks before I arrived at the station all the shifts were briefed on how to behave. The first couple of hours after I got there, nobody spoke. And then someone burnt his hand on the stove and swore. Everyone just froze. I was digging my hands in the seat, really nervous not knowing what to do. Then he said, 'What! We're not allowed to swear around her!' So I made a point of swearing a bit as a way to be accepted. But that's a long time ago now. Today I wouldn't be doing anything else. It's the camaraderie that makes this job so great. You're together! It's more like a family. It's not like an office where you go in, take off at lunch time with your friends and come back. You're locked in. When you go somewhere you go as a group, so you work at the relationships.

There can be a lot of downtime in this job so you have to learn to be patient, but when the time comes, you have to be able to handle the pressure. It's an odd job when you really think about it. We're asleep at 3 o'clock in the morning, then a few minutes later, we're pulling a total stranger out of a car accident, or walking into a house fire.

I really admire the people I work with. It takes someone special to be in the MFB.

DAVE KELLY

STATION OFFICER
43 YEARS OLD

I virtually came straight out of the army and into the MFB about eighteen years ago. I had a father and brother who were policemen and that quite interested me. I weighed up my options and I liked the type of work the MFB did, so I thought I'd give it a go and was lucky enough to get in.

One fire that sticks in my mind was one at the Wesley College. When still at the station we were getting ready when one of the guys slipped and caught the top of his head on the underside of a truck door. He tore the top of his skull open and knocked himself out. He was pretty badly hurt.

I remember driving down there, and as we got closer and closer we could see the glow over the tops of the buildings. We were first on scene, and decided we had to chop the fire off at a certain section. We went to work from there and after about an hour we cut the fire off, and in doing so had stopped it moving into the library, which was one of the most valuable parts of the school. It's always great to know you have made a difference in whatever way.

You don't seek out any congratulations. Occasionally you might go to the smallest little job where people are so grateful for what you've done. We sometimes have people write nice letters and cards, or they come down to the station with a slab of beer to thank us. We normally don't accept it: 'it's part of our job'.

Another one I remember is a house fire at about 2 o'clock in the morning. We rolled up to a single-fronted weatherboard with a lot of smoke and a bit of flame coming out. One of the neighbours said that she hadn't seen the man who lived there. We put our breathing apparatus (BA) on and went in. We virtually stumbled on top of the missing man as soon as we walked into the hallway. We were lucky. We dragged him out, and worked on him until the ambulance came. He lived.

When I was younger, death didn't even enter my mind. I used to think that if I went to a fatality I couldn't put a name to a face. I simply did the job and went home. But now as I've got older, my view has changed. There was a recent hanging we went to that hit home. We were talking to the ambos, taking a few details, and I looked at the date of birth of the victim. It was seven days from my birthdate; we were the same age.

Of the memorable fires I've been to I could probably tell you every person who was on the truck at the time. You always have the funny little stories about some of the fires you go to, or the fun that people make of you during the downtime.

One night I was in the watch-room at the old station at Windsor. The alarms for the area used to come into the watch-room and it was our job to monitor them. When an alarm went off, you would write down the details, get the keys, alert the crews by putting on the bells, and off you'd go. After 11 o'clock you were allowed to relax and recline if there was no work to be done. On this night I was in bed when I heard a tapping on the window. There was a pub about 100 metres up the road, so I thought it was one of the drunks. I pulled up the blind but no one was there, so I went back to bed. About ten minutes later, it happened again. *Tap, tap, tap.*

I ended up opening the manual doors at the station and looking out on the street, but there was no one. By then I'd lost my sense of humour. As soon as I got back inside, there was the tapping again, so I thought 'right you buggers!' I went outside again, then pretended to close the doors and go back inside. The tapping started again, I looked out and three of the boys on station had joined four broomhandles together and were leaning out the window upstairs tapping on the window just to give me the shits! I was actually surprised that when I went outside a bucket of water didn't come over the top of me because that was a common one.

Little things like that bring you all together. The best part of being in this job is the amount of friends I've made, all the good people that I've met. We have a special bond. I'm very lucky to be a firefighter.

DARREN HAY

LEADING FIREFIGHTER
39 YEARS OLD

I was an electrician before I joined the MFB. Sixteen-and-a-half years down the track I still enjoy coming to work. In this job you never know what's going to happen. Personally I don't want fires and accidents, but if they do happen, then you have the chance to do something about it. You save a life or save someone's property or possessions and it's a really good feeling.

Most of my biggest memories are about fatalities; I think most firefighters have a few terrible moments that really stick in their minds. One of my worst jobs was a car accident involving an elderly couple. The husband was in the driver's seat. He was on a steep driveway and, unbeknownst to him, his wife had walked around behind the car at the same time he'd decided to back down the driveway. She went over, he heard her scream, and in the confusion he went for the brake but hit the accelerator instead. She was caught under the car, dragged down the concrete driveway, across the footpath, straight backwards across the street and into the side of another car parked on the other side of the road.

We were first on scene. By the time we got there some blokes from a panel-beater's shop had come and jacked up the front and the back of the car, and the old woman was still lying under it. We checked to see if she had a pulse and if she was breathing. We didn't want to move her too quick in case of spinal injuries. She had a very, very faint pulse and her breathing was extremely shallow and ended up going into a reflex action where she wasn't actually getting any air in. We made the decision to move her as steady as we could. At that point the ambulance turned up. They agreed with what we were going to do. I climbed under the car and put my hands under the women's head to support it. My fingers actually went inside her skull but I just had to bear it. She also had a gash on her leg that was cut right down to the bone. There were multiple other injuries. We cut her clothing off, put defib [defribrillator] pads on her, and started doing CPR. There was that much damage to her skull there was no hope of saving her.

Once she was pronounced dead her husband was cradling her in his arms. They'd been married for about 50 years; I heard from a neighbour that they actually doted on each other. You don't forget things like that. I haven't cried from anything in this job yet, but I have a while to go yet, and you never know.

After that job we went back to the station and had a yak about it. Our senior station officer was really good; he spoke to us all and asked if everyone was okay with it.

In cases like that it's made known that if you do need to speak to anyone about it, the counselling facilities are there to deal with it. Personally I've never used it; I've never actually rung up the peer support group, but everyone is different.

On the other side you've got the rewarding moments. I remember one house we went to out in the suburbs, where there was one bedroom on fire, and the place was right on the edge of going up completely. I was actually in on my own because the station only had the one truck and was waiting for back-up from other stations. I had a choice: I could stand on the outside and wait for the next truck to get there, or go in on my own for a couple of minutes. I chose to go in. That was a bit scary for a while. There was a bit of flashover. [A flashover occurs when unburnt vapours—either vapours from the incomplete combustion of the fire or from materials in the room—each reach a point where they flash or explode. This usually happens across the ceiling where the vapours have gathered. This may cause some of the combustibles in the room to ignite. Sometimes the flashover stops until the vapours build up again.] I managed to hold it back, then beat it back into the bedroom and we then put it out. The people who lived there were stoked; they were absolutely over the moon and had nothing but praise for us. That made me feel good. Sure they lost a bedroom but the majority of the other stuff was safe. And for people it's not so much the house but the sentimental things that mean the most. The photos, the gifts, the things you pick up on your travels. So when you can save things like that it's really satisfying.

This job makes me realise what a loose grip we have on life. If I go to a car accident and see someone dead through no fault of their own I think, 'Well, at 8 o'clock this morning that person probably got up and was walking around like I was. He was just getting ready for the day ahead.'

Then bang, he's hit by a drunk driver or whatever, and he's gone. We go to the accident, and we don't know him from a bar of soap, but the rest of his family is sitting at home totally oblivious to the fact that their loved one is no longer with them. It's hard not to think about that. One thing is for sure . . . this job has made me a lot more safety conscious and appreciative of what I have.

SHANE TORR

LEADING FIREFIGHTER
43 YEARS OLD

I'm a maintenance fitter by trade. I did my trade for nine years, but began seeing that a lot of things were being made overseas and being imported, so I thought it was time to move on. My uncle was an ambulance officer, so I thought of doing that, but then the MFB advertised for positions. I didn't know anything about the service; I didn't know what the shifts were, but I knew it was a job with some security. That was nineteen years ago now.

Back then, there were more fires, probably because fire protection wasn't around as much, and we weren't educated as much as we are today. Nineteen years ago if you had a car accident there were times when you had to wait, but now with the mobile phone, 60 seconds later emergency services can be on the road.

I don't think a lot of people really know what we do. The first call-out I can remember going to was an accident where a guy got hit by a train. We had to go and find body parts in long grass at 4 o'clock in the morning because they didn't want dogs coming along, picking up a hand or a finger and taking it into their backyard. That was within my first year of joining the job. That will always be in my memory. It's not as though you think about it at the time because you are focused on the job, but afterwards you come back and think: 'Shit that was a bit bloody!'

The only other ones over the years I really remember are the ones when I've made a mistake or I've come back to the station and the guys have laughed at me. Like the day I put up the platform at a building fire. The problem was the fire was reflecting onto another building and I pointed the wrong way, thinking: 'Shit there's a fire next door!' I copped a lot when I came back and I will never live it down, but you've just got to learn to brush it off.

I'm happy to be here. It doesn't mean we are all the same, but we get on, and when we are happy and get on, we work together and trust each other. It's like when we have roast days. They're like a team-building exercise. Everyone gets together, and they talk, laugh, pick on each other. It's all really important to make the environment a good one that people want to be in. Above all, the MFB is a family.

PETER GILL

LEADING FIREFIGHTER
53 YEARS OLD

I joined the MFB in 1979. I had been working for an insurance company and wanted a change. It was totally different back then. The junior boy was the junior boy. He made the tea, got the lunches, did most of the cleaning in a small station. And there was more emphasis on rank; the officers didn't associate with the men as much and, of course, there was different equipment.

On my very first night, we got a big job. It was about ten past six and we were just sitting down to eat dinner, but didn't get back to the station until about a quarter past nine the next morning. It was a warehouse full of paper. I stood at the end of a hose with the BA on basically all night. I was wet and cold and thought: 'What have I got myself into?'

I also remember attending a decapitation in a train accident. When I went home that night and my wife put dinner on the table, I just said: 'Sorry, I can't eat that!' It took me a while to get over it.

I've seen a fair bit of death, like many of us here, especially the older blokes. You often handle it by joking about it afterwards. Seeing dead adults doesn't affect me as much as children. Young kids can really get you going.

I don't normally talk about work at home. I do when my wife asks me what's wrong, but you don't want to involve your family. You have to protect them too.

The MFB has obviously been a big part of my life. I'm coming towards the end of it now. And although there are things I definitely won't miss, like night shift, I'm thankful I've done a job that few others get the opportunity to do.

ANJA DOBAJ

QUALIFIED LEADING FIREFIGHTER
37 YEARS OLD

I migrated to Australia about seven years ago from Germany. I started to work in my profession as a plumber and gas fitter but wasn't happy, so I looked for something else. When the MFB was recruiting I did not have to think. I got in on my first attempt about three-and-a-half years ago.

The best part of the job is the companionship. Through my working life all my jobs have been with males, so it was no problem adjusting to this. For me, the gender question is not an issue. It's excellent, we have our fun, and it's good being one of the boys, although that's not the official term. The major thing is to look after your mate, especially when you come back from tough jobs where people have died or been injured. You talk together and we support each other.

I do feel for victims, but I know those emotions have to go away. I cannot really take part in a victim's or family's suffering. I have to be able to cope with it. I do go home and talk about it, but I did in my other jobs as well. I'll go home and swear about the bad days, and laugh about the good ones. I spit it out, then it's gone. I am lucky I can switch off.

My favourite role is definitely going into a fire. Grabbing the gear, going in, and hopefully saving properties and lives. The action bit. It makes your heart pump faster, but you aren't worried because you are well trained. It's not as though there are millions of things to think about when the tones go on. Once they do, your reaction becomes second nature.

I've never had to drag anyone out dead or alive, but I have had to throw a cat out a first-floor window!

I've had plenty of car fires, which are pretty good because they create a lot of smoke and get you used to your BA and all the stuff you learn in the training college. The more exciting bit was my first house fire. I'd never seen that many flames! It was at night time and pretty well across from where I live, so I was thinking: 'Oh, this does happen to anyone.' It was well alight; the good thing was that it was an abandoned house, but then we found a mattress at the back of the house. My heart pumped a bit because it would have been my first victim or rescue. Thankfully there was no one inside, but that incident made me alert as to how important our job is. You never know when you will be needed to save a life.

GREG PLIER

QUALIFIED FIREFIGHTER
28 YEARS OLD

The first time I thought about becoming a firefighter was when I was about seventeen.

I ended up going to TAFE to do a fire technology course, and I followed it from there. I ended up applying four times over about a five-year period to get into the MFB. I even tried in other states, so it was nice to finally get the call: 'You're in!' That was three-and-a-half years ago.

The high points of being a firefighter are the feelings you get after doing a good job, even if it's a small fire.

I suppose one of the lows is how quiet it can be, especially in town because there's not as many fires as you might get in the outer suburbs. Not that you want fires, but when you're trained to do a job and you don't get to do it, sometimes you think: 'Oh gee!'

One of my earliest memories is of the first week I drove the truck. We got a call to a house fire, which was pretty intense. I remember speeding along, cars all over the road, listening to the radio, and my heart racing. You get used to it, but especially around the city you have to be pretty wary. You might think that a lot of people pull over to the side of the road and let you pass, but not many do that. They'll pull out in front of you and want to follow you through red lights. Taxi drivers are the worst, they're shocking!

This job definitely gives you a different view on life. I sort of normalise death a bit because it's just what I do for a job, but when I stand back and look at it I realise the effects it can have, especially incidents involving kids, because I've got kids.

Overall though, it's a great job and I don't think there'd be many people in the service who'd say they want to be doing something else.

IAIN ARGUS

FIREFIGHTER
38 YEARS OLD

I used to work for myself as a plumber, and I got to the stage where I wanted a career change. I looked at the MFB as an option about twelve years ago; a lot of guys I used to play footy with were in the job and they encouraged me to join. I finally got accepted in April 2003.

For most of the younger guys in the job—I'm 38 but still am one of the younger ones in terms of experience—the biggest thrill is going and fighting fires. You get that hit of adrenaline. It might be the middle of the night, you're sleeping in your jocks, T-shirt and socks, then the tones go on. All of a sudden you're at the engine bay getting your gear, and all the time your heart is just beating through your chest. You feel it in the truck when you're listening to the radio, hearing the siren. And then you get to the job and another hit of adrenaline gets you even before the job. My first fire at night was a car fire. When we got back to the station and went to bed it took me about three hours to get back to sleep because my nerves were still racing and I was re-living every moment. It was huge. It was like playing football. You play finals and big games, they didn't even compare to the adrenaline hit I got that night. I've jumped out of a plane and that didn't even compare; it was nothing!

There's no doubt this job has given me a different view on life. I'm now more aware of the people around me. If an accident happened in the street, I know I can help, whereas before I might have turned a blind eye and waited for someone else. That sense of responsibility and knowing you can be a service to the community is very satisfying

I'm definitely a better person for being in the MFB. I'm more relaxed and have a better outlook on life.

ERIC NOBLE

LEADING FIREFIGHTER
46 YEARS OLD

I've been in the MFB for eighteen-and-a-half years, the last six here at South Melbourne. I'd been a landscape gardener and a mad surfer, but hadn't thought much about a career because I was too busy throwing a board and a tent in the car and going off doing my own thing. I was 27 when I joined, and at first I didn't really have a sense of doing a service for the community, but that soon developed.

It's the big jobs that stick with you. They give you a buzz. It's just a natural thing, isn't it? It's the body's and mind's reaction to a pretty unnatural environment. I still get a buzz out of turning out to a big fire, even when I'm driving. The actual process of firefighting is the best part of the job; going to the scene, running a line into a pump, getting the pump going, running a firefighting line and actually putting a fire out. Actually being part of it. It gives a real sense of satisfaction, teamwork and self-worth.

Being inside doesn't happen all that often. I could count on one hand the times that I've done it, the times I've stood in a doorway and watched flames come out. But it is good, the adrenaline takes over. I don't know whether it happens to the guys in the army when they're under attack. You just do it. Your actions become automatic. It's exciting.

I remember a lot of car accidents where people have been injured or killed. We don't see nearly as much of it as the police or the ambos, but the ones you do tend to see, stick. I could probably talk about every fatality in detail or every bad car accident that I've been to.

There are some that really stay with me. A lady had locked herself in the family garage, tipped petrol over herself and set herself alight. The husband had just gone and dropped the children off to school and he drove back down the street to find all these emergency vehicles there. I can remember afterwards that the husband actually thanked us for doing a 'really good job'. He was in shock, and that shocked me too.

There was another one where a guy died after an explosion in his greenhouse. It was a cold, foggy drizzly morning, and there was a young policewoman who was told not to let anyone near the man's bloated burnt body because it might have been a crime scene. She had tears rolling down her face. I felt so sorry for her. None of these jobs are easy are they?

I know you would have been told this before but it's important to understand how important black humour is to us. There have been times when we get back from a horrible job and the blokes might break the ice with a really sick joke. You think: 'Oh, you can't say that,' but then it helps us to deal with it, and unless you've been there you wouldn't understand. I talk about it when I go home, and when I am telling someone I feel as if I'm really hard-hearted, as if it's black and white. But the fact is that you didn't know that person or their family. You need to distance yourself from it or this would be an impossible job to do.

In the case of accidents, I do sometimes wonder about how the people we have helped got on. Sometimes we'll find out by talking to an ambo or a policeman on the next job, but not often. After an accident it's important to move on.

The best part is the people you work with. You do come in and have a laugh, and share so many experiences. And of course there's the excitement of going to a job. Everyone looks forward to going to a job.

ROWAN CHAPPLE

LEADING FIREFIGHTER
44 YEARS OLD

I used to be in concreting, but when I hit about 29 I knew I couldn't keep doing it forever. I was lucky enough to get into the MFB first go. I'm now in my sixteenth year, and I love it. I love the social part of it, the mateship. You're living with people for short but intense periods every week. You get pretty close to people and if you don't you're not going to find it easy.

Over the years I've done all sorts of things. I've had to climb trees to get cats out, and even ripped walls apart where a cat has been caught in a cavity. We had a kid one day who got his head stuck between some stair-rails, and then there was the lady who used to get drunk and lock herself out of her house. She used to ring up and say: 'I think I've left something on the stove.' We'd go there, break her door down, and in she would go: 'Thanks guys, see you later.' She knew the MFB would turn up and check her report. She was pretty wise to it.

I'd be a liar if I said I didn't feel fear or apprehension at times. I certainly remember going to one hotel fire. I walked into the building with my officer and there was quite a bit of flame going around. We walked underneath some stairs into a kitchen and the first thing we could smell was gas. My officer said, 'Take note of that.' We then looked at the jets on the stove, which were all fully on. The officer didn't muck about: 'Let's get out of here!' We went straight out. We found out later that the stairs collapsed about two minutes after we'd walked out. That was probably the scariest moment I've had. I certainly look back at that and think 'what if?'

You have a different sense of urgency when you arrive at a job and are told there could be people inside. I look back on one house fire where I had to go in and search for a boy. I couldn't see my hand in front of my face, and was looking under beds, in toy boxes, under tables, in every corner of every room. We searched the whole place, hoping that we didn't find anyone. Fortunately the kid had been next door. I did a similar search at a brothel, but could only check one room before having to back out. Looking for people is one of the worst parts of the job, and not knowing what you'll find.

But you do have the good moments that can happen even in the worst of accidents. I have come away from a lot of car accidents feeling pretty good because we've helped people out. That can be a great feeling. Occasionally you get someone who will say, 'Thanks very much,' but not often. That doesn't worry me. As far as I'm concerned I get paid to do a job.

What does being a firefighter mean to me? Well, you have pride in who you are. People tend to look up to you. That is really nice.

CRAIG SIMPSON

LEADING FIREFIGHTER
46 YEARS OLD

I joined eighteen years ago. I always wanted to get into an emergency service of some sort. My history was surf lifesaving culminating in the helicopter rescue division for my last few years.

I visited a fire station as a kid, and I still remember like it was yesterday. I thought the blokes were sort of 'knights in shining armour'. They seemed to be going into dangerous environments and putting themselves at risk to serve the community. They meant a lot to me.

There's this joke about firefighters that sums us up pretty well. The story goes there is this researcher doing a thesis on human behaviour, and he decides to get three people from three very different jobs. He gets a scientist, an architect, and a firefighter. He locks each of them alone in a room for 30 minutes with two ball bearings. He comes back to find the scientist has one of the ball bearings orbiting around the other one, and the architect has one ball bearing mounted on the other as though they are a building. He then goes to the firefighter who says a bit sheepishly: 'I've lost one of the ball bearings and broken the other one.' I think all of us have lost and broken things over the years!

Fire is a strange beast. It can be frightening but it can also be very relaxing. You can be in front of an open camp fire watching the flames and it can be comforting but there's probably little more frightening than going into a raging out-of-control fire. We like nothing better than a big burn. No matter how many fires you attend, the heart still thumps. If people don't say they're a bit concerned or frightened, then they're kidding themselves.

One job I had was at a single-storey weatherboard place. It was mid-afternoon. We were first truck on scene and there were flames coming out of all the windows at the front, and there was smoke coming out the eaves. There was a gentleman reported in there who was a severe asthmatic. When someone isn't accounted for you take a few more calculated risks than you would if there was no one in the building. Two of us went in and when we opened the door, straight away there was black smoke rolling out. We went down on the floor. At this time the guys were just getting the hoses and feed lines into the pump. The hoses were brought up to the door. My buddy and I were both in BA; he grabbed the hose line and started hitting the fire. From the floor it just looked like a big storm cloud, black and rolling. About a foot and a half off the floor it was fairly clear, and a really bright orange colour like someone had a big torch. I couldn't see the fire at this stage, and we were into the building. We eventually saw the guy under a table in the kitchen. Due to the build-up of heat, things must have flashed over and we could see the flames coming through the smoke. Then the windows imploded and let in a whole lot more air, and caused your genuine backdraft. The flames just shot out right over our heads through the door behind us and ignited the wall of the place next door. I still have the helmet I was wearing at home. It is all scorched. We eventually got this bloke out. He was huge, about twenty stone plus. We resuscitated him, but he died about five days later from internal burns. That was probably the hairiest moment I've been involved in.

I guess you have to think of fire as the enemy. The big jobs are obviously dangerous, but then again you can come just as unstuck in a simple garage fire with a bit of brake fluid or pool chlorine or something like that. You walk in the door, a can of two-stroke goes bang right beside your head and that's it!

Being a firefighter has made me appreciate what I have. Not only when I go to incidents where people are injured or have died, but I see some people living in squalor, or they're a victim of circumstances; car accidents are a classic example where an innocent person can be in the wrong place at the wrong time and their life is suddenly over. I really feel for those people.

I very much separate life here from home to the point where I don't really talk about it much. If something happens that has been a bit stressful my wife will pick up on it but she won't push for answers.

You don't have to look far to see the brotherhood that exists among firefighters. If something happened to one of us, there'd be a bunch of people you could ring up and say 'listen I'm in a bit of strife could you give me a hand'. They'd be there straight away. It's pretty close.

After all these years I still enjoy being operational. There is so much to do and so much to learn that the responsibility is on us to be able to help out at whatever emergency we have to, and that's a big responsibility to be given.

STEPHANE VICTOR

LEADING FIREFIGHTER
36 YEARS OLD

I joined in June 2000. I had been a gym manager, but wanted to do something that was community-minded. Do a bit of hard work and save a life or two was something that appealed to me.

The highs are always the mates you make; you get to knock around with some pretty good people.

One of the hardest things I've done is to turn out to a man who'd decided to douse himself in petrol and set himself alight. He was a refugee on a temporary protection visa.

It made me realise how lucky we are in this country in so many ways.

I am no different from a lot of people in the job. The notion of actually turning out to the bad accidents and fatalities in itself has a psychological protecting factor. If you were just to come upon a fatality as you were driving home from work it would be a different thing, but the fact it's part of our job makes it easier to deal with. That, and the fact that you're with a lot of people to support you. You always have a strong network to draw on. Without it, the job would be much harder.

Although it might sound strange, working in this type of role brings out the best in people. Although you are generally attending someone's misfortune, you see the strength of humanity. I'm a humanitarian-oriented person. Just to work with other people and learn from them, while developing a sense of camaraderie has been a fantastic experience. It certainly has made me realise that when you're in a pinch there are a lot of people who will go to bat for you. That is really heartening to see on a daily basis. Being a firefighter is what I get out of bed for.

JULIAN BISBAL

FIREFIGHTER
31 YEARS OLD

I've been a firefighter for two-and-a-half years. I was a plumber before that. A lot of firefighters play down what they do but we're little kids at heart and we love it.

I can remember my first few fires because adrenaline kicked in, and I was a bit scared too. The training teaches you to put a little bit of water away and wait for the steam to help put the fire out. The experienced guys are always telling us: 'Don't use the water until you see the red stuff.' But I just kept on going and going until I felt the steam come down and it burnt my ears. I was putting too much water on, then not enough, too much water on, then not enough. It took me a while to get into the swing of things, and now I'm watching the other guys coming through making the same mistakes.

I remember one fire in a toy factory. They were telling me there was a forklift in among the flames which had an LPG container in it, but they didn't know where. I was standing there with this 50-mm line thinking: 'Okay just be careful', and sure enough there was this massive pop and the next thing you know this lid from a 44-gallon drum had landed right next to me. One of the more senior firefighters grabbed me, pulled me back by the tunic and said, 'Don't move.' And I said, 'Mate, I am not going anywhere!'

There was also a car fire I went to where a guy had set himself alight. His father was on the scene within minutes, and to see him pull the sheet back off his son's face will stay with me forever. There are some moments you know you won't forget. On those occasions you talk among your mates, and nine times out of ten you can reason why it happened. You have to tell yourself that you're there for the good of the people, but you can't always save everybody.

Fighting fires can be amazing. From the moment you're in the truck you generally know what type of incident you're going to. And if you know it's a good factory fire or a house fire and it's in the middle of the night, then you start thinking, 'Okay this is a big job.' You're doing your collar and tunic up a bit tighter, you're making sure your gear is ready, your gloves, your face mask, everything you'll need.

When you go inside on an internal attack, you're looking for the ceiling, wondering, 'Is it going to fall down on me? It's weird, it's almost as thrilling as it is scary. I know they are opposites but they go hand in hand. Sometimes, even though you know there are people around, you can feel very alone. You can't see two foot in front of you. I've been to jobs where a senior guy has said: 'Put your hand on my shoulder and come in with me,' and I couldn't see past my elbow. I knew he was there but I still felt pretty alone.

Fire is an amazing animal. It's volatile and alive. It's absolutely mesmerising but you can't turn your back on it. You always have to respect it because it will beat you if you don't know what to do.

Being a firefighter has taught me not to assume things. I used to judge people a bit more than I do now. I'm learning more and more every day just by talking to people in the job. No one should take anything for granted because you just never know when something might turn against you. Love your kids, love your family, and love your friends.

COLIN AND ROBYN HETERICK

'Married to the Job'

> 'The best thing we have in this job is friendship. All the friendships we've made over the years, and the friendship that Robyn and I have.'
>
> Station Officer Colin Heterick, Queensland Ambulance Service

Ambulance Officer Colin Heterick liked to arrive early before a shift. His ritual of sitting quietly for a few minutes allowed him to prepare for the hectic hours ahead. Once he drove out the doors of Sydney's Quay Street Station, he knew he'd often not return until it was time to clock off. On any given day, or night, they could be called out to: a cardiac arrest victim in a multi-million-dollar office; a diabetic in a tiny dilapidated unit in Redfern; a heroin overdose victim in a back street of Kings Cross; a cocaine overdose victim in a plush apartment with harbour views; a homeless alcoholic; a baby with asthma in fashionable Vaucluse . . . life was never dull.

Since he'd first become an ambulance officer in 1988, Colin had learnt that nothing should ever be assumed or expected. Uncertainty was a central theme in his job, and so it was the day his station officer approached him when he was alone with his thoughts before a shift.

'I want you to work with someone new,' said the station officer.

'Who?'

'Just someone who's finding it a little difficult to find her way around the city. I just need someone to help her, and you know your way around.'

'Alright,' said Colin, a little reluctantly. 'So who is it?'

'Robyn Ritchie.'

It was late 1990 and Robyn had only been at Quay Street for about six months. She was a quiet woman with a shy smile and softness of manner. She was 29 years old, and a mother of three children who was taking fresh steps in her life after separating from her husband. She had been in the New South Wales Ambulance Service for less than a year; the challenge of adapting to such a demanding career was made even more difficult by her 90-minute train trip to and from Sydney every day from the Central Coast city of Gosford. Considering day shifts were ten hours long, and nights an extra four hours, Robyn had little time for anything other than work, travel and sleep.

In contrast to his impending new partner, Colin wasn't afraid to speak his mind, even if it occasionally incurred the wrath of his superiors. He was 21, the age at which many men believed they knew everything, only to find themselves, years later, saying: 'If only I knew then what I do now.' Nevertheless, Colin was street-wise, and had a bit of self-confessed 'mongrel' in him. After spending his first few years of childhood in Sydney's western suburbs, he moved with his family to a number of smaller towns near Orange in the state's central west. He left school when he was fifteen, and headed to Wagga where he started a fitter and turner's apprenticeship with the Royal Australian Air Force. He didn't finish, and eventually returned to school in Orange to complete his senior studies and also to 'have a bludge, and play more sport'. After closing the textbooks again, he floated from job to job, including pumping petrol, moving furniture and picking fruit, the last of which was much more orderly than the mischievous days of his youth when he and his mates sometimes 'accidentally jumped the fence, accidentally shook the trees, accidentally held our arms out, and accidentally ran away'.

When octane, tables and apples gave way to contemplation and more maturity, Colin seriously considered his future. The thought of becoming an ambulance officer appealed to him, and so he applied. He already had a reasonable understanding of an officer's lot because his mother had fractured her leg in a car accident years earlier, and ever since then she'd been in an ambulance many times on her way to hospital for rehabilitation. Colin enjoyed talking to the various officers who came to the Hetericks' front door, and his interest eventually grew into enthusiasm and desire.

After completing five weeks of training at the New South Wales Ambulance Service Education Centre he did a probationary period at Blacktown, in Sydney's west, where the pace and diversity of jobs made him realise 'it's great to have all the theory but you still know bugger-all until you're out in the thick of it, and learning about yourself in all different types of situations.' That education continued when he transferred to Quay Street about six months before Robyn arrived. When his station officer told him that he was to have a new partner, Colin approached Robyn with the news.

'Did you know we're going to be working together?' he said

'Oh no! We're not are we!'

Until then, the two had only met in passing. They'd worked on different shifts, and really only knew each other by reputation, something which didn't sit comfortably with Robyn, who'd heard from others that Colin could be a bit of a 'hot-head'.

On their first shift together, they both arrived early for an 8 am start. Robyn soothed herself with a cup of tea, but her nerves quickly tightened when the PA announced a call-out.

'You're driving,' said Colin, tossing the keys at his partner.

'No, I'm not!'

'Yes, you are. And you'd better hurry up because we've got a job to do!'

Colin walked calmly around to the passenger door of the Ford F250 and got in.

The keys soon jangled in the ignition switch, and Robyn edged forwards out of the station doors.

'Which way do we go?' she asked.

'Well, you can't turn right, so you better go left!'

Colin couldn't hide his smirk. If this anxious woman sitting next to him was going to learn the ins and outs of the city's geography, there was only one way to be taught . . . behind the steering wheel. It was a fast-track way to learn but hopefully not a crash course.

Over the following weeks, Colin rarely drove unless extreme urgency demanded it. On those occasions, he was driven by adrenaline and, even now, more than a decade later, he grins when he remembers such times.

'I've jumped off bridges, I've gone on jet-boat rides, I've been on roller-coasters—I won't jump out of planes mind you—but there is no greater rush than putting the lights and sirens on and flying down George Street. Whenever you put those lights and sirens on your adrenaline goes up.'

But Robyn felt only apprehension until she was comfortable enough to know that the Queen Victoria Building was opposite the Town Hall which was one block up from the Hoyts Cinema Complex, which was just a right-hand turn away from Chinatown, which was a straight drive from the Sydney Entertainment Centre, which was a U-turn, short straight, a right, another short straight and a left from Central Railway Station, which was a . . .

It took her about a month before she felt she had a reasonable knowledge of the city's veins and arteries. She became more relaxed and confident about both her driving and clinical ability, and all the while Colin kept his mongrel attitude on a short chain, and surprised his partner by being supportive, quite literally at every turn. Robyn began to wonder. Perhaps this apparently brash young man with the wavy brown hair and athletic frame isn't at all as his reputation suggests?

Their days soon shifted into nights, where work often began with inter-hospital patient transfers. Then, if time allowed, it was a quick stop for dinner before the call-outs whipped up the speed with chest pains, shortness of breath, alcoholic cases, epileptic fits, occasional car accidents, household injuries and, of course, the predictable happenings in the backstreets of Kings Cross where a junkie's mattress and pillow was the gutter and footpath. There were also the times when crews were called into the peep-and-porn establishments, where pink neon naked ladies leapt across the entrances behind which flesh illuminated imaginations in the darkness.

Just a few nights after first attending to a heroin overdose victim together in a laneway, Colin and Robyn were called to another victim in a well-known nightclub. They walked through the main bar area up a flight of stairs and along a hallway past some more bars and a number of rooms with doors shut. Colin kept going. There was no need to stop, stare and wonder about the tools that adorned the walls, nor did he have to check stride to look at the women, many of whom were his age, whose attire was somewhat less formal than his ironed shirt, trousers and polished boots. But as Colin strode, Robyn stuttered. A child gazing at a falling star couldn't have had wider eyes or a mouth more open. Welcome to Sydney's hidden galaxy of sex workers.

In the days afterwards, Colin found himself thinking about the incident. He'd seen it many times before: the seedy joint, the hookers, the chains, the whips, the trip gone wrong, but yet he'd walked out feeling refreshed and quietly excited. Certainly it was always rewarding to successfully treat a patient—in this case he and Robyn had oxygenated the victim and waited for the more qualified Intensive Care officers to arrive—but there was something else that had given him an eagerness to keep moving. Since he and Robyn started working together six weeks earlier, he had come to know a cheerful and amusingly naive woman whose character partially reflected her upbringing. She was initially a country girl, born in the northern New South Wales town Inverell, before moving with her family to Brisbane. She married young, and had two sons and a daughter by the time she was 22. Yes, she had grown up in a hurry, but in many ways she was still to live. Her freshness was appealing, so too was her broad smile and bright blue eyes. Colin was beginning to think, 'There could be something more here if I work on this properly!'

His project began with little acts of kindness that could have easily been dismissed as nothing other than the products of friendliness and good manners, but behind these was a plan. When Robyn arrived at the start of each shift, she was met with a cup of tea and a boyish grin. The strategy further developed at one of the few jobs that Robyn was bored by . . . the Friday-night trots at Harold Park. Colin came prepared with some of his partner's favourite music, and as the pacers flashed down the straight with hooves pointing, the ambulance trailed with its crew's toes tapping to the tunes of Fleetwood Mac and Meatloaf. While this entertainment ploy was successful, more still needed to be tried, and as with most men hoping to impress, Colin sought a dinner location that would melt good food and atmosphere together in one sensitive package. However, he faced the problem of being drawn away when just a mouthful into the meal, so everything had to be adaptable. His choice proved perfect . . . a takeaway Thai restaurant at Kings Cross with the best satay chicken sticks in the city. Robyn was impressed.

But a mere few blocks away from such delight were people who could make Colin's quest more difficult, and perhaps embarrassing beyond repair. They stood on street corners near some of the most prestigious car dealerships in Sydney. They had short skirts, skin-tight tops, high-heels, sheer stockings, immaculate hair, long nails, enough make-up to paint a canvas, and . . . they all knew Colin.

It wasn't long after the nightclub incident that Colin stopped at a set of traffic lights and was immediately swarmed on by a group tottering out of the shadows. One member pulled out some lipstick and launched into what seemed a well-versed routine between exaggerated puckers.

'Oh, lovie, can I borrow your mirror? Don't you look so strong and manly in that uniform. Mmmh! After you finish your shift why don't you come back here and we'll have a good time. Wouldn't that be wonderful! How about it sweetie?'

Colin tried to ignore the request, which prompted Robyn to say with the utmost innocence, 'She's talking to you. Why don't you answer her?'

'Yes, I know, it might just be good to shut up though, please,' he said, hoping the lights would change smartly to green.

In the meantime, others in the group were looking at their reflections in the ambulance windows and making the appropriate adjustments, from the plucking of an eyebrow to a lick of the finger and a quick pat down of errant hair.

When the lights eventually changed, Colin, with much relief, accelerated away, leaving the mock-smoochers to return to their corners and prepare for their next display of street theatre.

'They're nice girls, aren't they?' suggested Robyn.

'Yeah, real nice GIRLS!'

'What do you mean? They were so sweet wanting to take you out to dinner.'

'Robyn, they were transvestites!'

'Oh, shit!'

Colin indeed had a good relationship with many trannies who were particularly helpful when an ambulance crew had to enter a dangerous area, such as a nightclub in which a fight had erupted. If a 'six foot six broad-shouldered woman in a skirt' wanted to escort him through the fracas, he never rejected the invitation. Robyn, too, would experience such moments where 'women became men again'. But on this, her very first brush with feminine Adam's apples, she copped the punches, albeit cheeky verbal ones from her colleague.

Robyn's naivety did create problems. After about two months working together, Colin thought all the cuppas, chook sticks, tunes, and cheerfulness surely must have given his partner some signals. Apparently not. Robyn continued to arrive at work each day without sensing that there was more being built than a good working relationship.

It took another colleague's quiet word to open her eyes.

'What! No! I don't think so,' Robyn replied to her female friend.

Others at the station had also assumed what was happening, but Robyn wasn't yet convinced. Colin was simply being a nice bloke and that was that. By this stage yet another strategy was about to be implemented. It was clever, it was appropriate and surely, surely, surely it would shake his partner awake. When Robyn received a carefully wrapped parcel from Colin at the start of a new shift, her eyebrows lifted.

'What's this?' she asked.

'It's a present.'

'Why?'

'Just open it and see.'

She unpeeled the paper to find a familiar tool of the trade.

'A stethoscope! What do I want another one of these for?'

Colin didn't say a word.

Robyn inspected her gift more closely. Oh! This was no standard-issue instrument, but a very expensive Littman's stethoscope, considered one of the best in the world. She turned it over, and cast her eyes along the steel strips that led to the earpieces. Ohh! She looked away, but the engraved words that she'd just seen were staring at her:

'To Robyn, love Colin.'

Ohhhhhh!

Robyn was quiet for the rest of the morning, prompting Colin to think, 'Oh shit! Have I made a mistake?' He'd already sought the advice of a colleague about work romances, and had been encouraged by the response: 'Go for it. As long as you can stay friends if it doesn't work out there's no harm done.' But now, he wasn't so sure. However, by the end of the shift Robyn was talking again. She thanked Colin for the gift, but stopped short of acknowledging that she knew what was happening: she was being chased. Over the following days they hid their thoughts behind stretchers, patients and paperwork, but this only added to Colin's anxiety. He needed to know where he stood. So, he decided he would make one final attempt with an action that couldn't be misinterpreted or forgotten about in the trials of a call-out.

He waited a few more days until he and Robyn returned to night shift, then one relatively quiet evening when on station, he launched his boldest plan.

'I don't know whether you're aware of it, but did you know there's a lot of wildlife around here. You know, possums, bats, birds,' he said to his partner.

'Where?' replied Robyn suspiciously.

'The Domain. How about we go and see the possums?'

'There are no possums in the Domain!'

'There are. How about we grab a bite, then I'll show you.'

Robyn's curiosity led her to the ambulance. 'Harry's Café de Wheels' was their first stop, the pie-selling institution at Woolloomoolloo that had thrown peas and potatoes in with the meat for generations. No doubt its workers had seen countless embraces and moments of high passion among its customers on the footpath, often in the morning's earliest drunken hours. But could they read a sober man's thoughts?

And if so, what was that ambo bloke thinking when he drove away well-fed but yet to satisfy a growing hunger to know?

Colin soon turned the ambulance into Art Gallery Road, which ran alongside the Domain, a favourite place for office-workers to relax on the grass or play a game of touch-footy in the lunchtime sun. But at night, it was deserted. Except for possums; or so one edgy ambulance officer hoped.

Colin pulled over and parked. He started talking. It was just small chatter about the day, the job, whatever topic flowed without effort. Robyn responded. Nothing seemed unusual. Here in the late night of a quiet shift, two workers were simply passing the time. Minutes went by, but no possums appeared. The conversation moved along politely without any mention of the stethoscope. Then, there was inexplicable silence, that uncomfortable moment when words fall into a black hole and parties are left to play the game as to who speaks next. But Colin wasn't in the mood. Now was the time. Motivated by the 'bugger this' approach that many a man can attest to, he leant across and kissed his partner. With a reaction that perhaps many a woman can attest to, Robyn thought: 'Oh shit! Oh shit! Oh shit!'

But she liked what had just happened . . .

Nine years later, Colin and Robyn were married in a private garden in Brisbane. It was a simple affair with 'two spanking VT Holden Statesmen', and a Woolworths-made chocolate mud cake that was shaped in two hearts. When the direction came to kiss the bride, Colin and Robyn were too busy murmuring to each other to hear until they realised everyone was looking at them.

'Is that it?' they whispered to the celebrant.

'Yes, you are now married.'

Nowadays, Mr and Mrs Heterick continue to work together in the Ambulance Service, but their journey has taken them far beyond the myriad experiences of Australia's biggest city. The days of seedy nightclubs have long gone, and the only trannies they now see are those that the locals use to listen to the stock and weather reports. They live and work at Taroom, a rural town of about 1500 people, 500 kilometres north-west of Brisbane. It has a small hospital, a doctor, a retirement village, and 18,000 square kilometres of potential dramas that ensure a challenging mix for an ambulance officer.

They arrived here in 1999, having moved north of the border six years earlier. At that time they were a solid couple, but the stresses of working at Quay Street had begun to stretch their relationship, so they sought a quieter life where they could spend more time together, as opposed to passing each other in the corridors between shifts and call-outs. After first looking unsuccessfully at options in country New South Wales, they responded to ads from the Queensland Ambulance Service, and were accepted at Proserpine, a sugar-growing centre on the far north coast in the Whitsundays tourist belt. They bought a house and stayed for two years before they split up and sought time for themselves. While Robyn initially stayed where she was, Colin headed to Winton, the remote town in the north-west outback that was famous for being the birthplace of Qantas. He remembers the period fondly.

'I loved the isolation and the friendliness. It was as far away from Sydney as you could possibly think. I'd often drive six hours to see a patient on these huge properties where they don't talk in acres but square miles. They had massive homesteads with surrounding verandahs and you'd swear you were in another world. I'd arrive and someone would want to sit me down, give me a cup of tea or a piece of strudel. I'd ask about the patient, but unless it was really urgent they'd normally say, "He's been waiting five hours, another five minutes won't hurt him. You've had a long trip."'

In the meantime Robyn moved to Brisbane. She and Colin remained in contact and eventually tried to take a step towards getting back together again. Robyn accepted a posting in Longreach, which was still a few hours' drive away from Winton, but certainly much closer than Brisbane was. This long-distance relationship gradually shortened the emotional distance between the two to the extent that Robyn delivered the ultimatum.

'Either we try again and make a commitment, or we split for good.'

Colin responded by moving to Longreach, where he and Robyn spent eighteen months. From there, they moved to Taroom. Colin was appointed Officer in Charge, and Robyn was his sole staff member. They were both Advanced Care Paramedics, and remain in those roles today.

Servicing a vast country shire has its own distinct challenges. Distance is obviously one, but it's others that take most getting used to. Although there are times of frenetic activity, the workload is generally low; it's possible to go days, sometimes weeks without attending to an incident. Official work hours are between 8 am and 4 pm on a ten-day cycle followed by four days off but, technically, both Colin and Robyn are on call 24 hours a day. One of them is always rostered on, meaning it's impossible to have time away together. They live next door to the station, and their mess is their lounge room. They are, quite simply, not only married to each other, but also their jobs.

'This is where it's so different from working in Sydney,' says Colin. 'At the end of a shift in the city, that's it, off you go and have a drink or forget about work for the day, but here you never really leave it. When we do get the chance to get away we get right away and we throw the bloody phones away! There's also the image you have to portray. I've been to the hotel here only a couple of times in seven years. In a small place like this you just can't be seen to be doing the wrong thing. Word gets around pretty quickly. We have our friends in town and we obviously know a lot of people, but it's so important that we have each other. And the great thing is that this job has given us a greater appreciation of our relationship.'

That appreciation comes from the experiences they share. As with most ambulance officers, they have vivid memories of particular cases. They have delivered babies—Colin has done so on the Sydney Harbour Bridge—but inevitably it's the tragedies that have the greatest impact. Colin can recall in intricate details every SIDS (Sudden Infant Death Syndrome) case that he has been involved with, while Robyn remembers the time she was in a vehicle with a dead baby and its parents as the tearful mother pleaded, 'Please don't let them take away my baby.'

Robyn cried too.

A limb twisted in farm machinery, a fatality in a car accident, a fractured skull from a horse-fall, a cowboy's leg pierced by a bull's horn, a drowned child, a burnt grandparent, a suicide... These aren't the usual dinner-table conversations between husband and wife, but the Hetericks understand the importance of discussing such moments. In their profession, they are together for better, for worse.

'We have talked about what happens on the job since very early on,' acknowledges Robyn. 'You need to for your own peace of mind, and when that other is your husband or wife, you've got additional reasons to do it. Colin and I are very lucky because we understand each other on the road, and that has helped our relationship as a couple. We seek advice from one another, but above all it's good just to know that you can tell someone who will listen.'

And perhaps a certain instrument has helped ensure that the Hetericks know better than most the values of listening. As Colin jokes:

'I gave Robyn the stethoscope so she could listen to her heart! Actually I didn't, but it makes for a good line now.'

Neither in this husband and wife team knows what line comes next. Importantly, they both enjoy interests away from their work. Robyn treats and cares for injured and orphaned possums, while Colin collects sporting memorabilia, especially cricket photos and models of Holden motor racing cars. Both are also studying through correspondence courses—Robyn, a Bachelor of Health Science degree; Colin, a Bachelor of Management—but the main lessons they learn are on the job where 'no two cases are ever the same'. Despite the unpredictable nature of their work they have enjoyed a constant for much of their professional lives. That is, they have had each other. And wherever, whenever they choose to leave Taroom, they will do so with much more certainty than their first left turn together out of Quay Street Station.

MICA Paramedic Mark Burns, Rural Ambulance Victoria. 'I'm a better person for doing this job simply because it's given me a better outlook on the community as a whole, and a better sense of self worth.'

Mark's family: Wife, Alison, with *(front from left)* Max, Anna and Will. 'As a parent I've now seen it from both sides, and that has given me a greater reason to think I'm doing the right job.'

Chief Superintendent Bernard Aust. 'The main intersections of your life determine who you are. I have been so lucky.'

Senior Firefighter Kevin 'Billy' Boyle (*centre*) with Camp Smokey kids, Kunthear, Emily and Rebecca. 'Just to see people smiling and laughing … makes me happy too.'

Dave Cuskelly (*far left*) at the time of his graduation from the Queensland Police Academy. 'Always be positive, confront negativity, and never assume anything.'

The MFB 38 C-Platoon (*back row, left to right*): Qualified Firefighter Iain Argus, Leading Firefighter Eric Noble, Leading Firefighter Darren Hay, Senior Station Officer Andrew O'Connell, Leading Firefighter Stephane Victor, Leading Firefighter Peter Gill, Station Officer Louise Cannon. (*Front row, left to right*) Leading Firefighter Craig Simpson, Leading Firefighter Rowan Chapple, Qualified Leading Firefighter Anja Dobaj, Qualified Firefighter Greg Plier, Firefighter Julian Bisbal. Absent: Station Officer Dave Kelly.

Chief Fire Investigator Danny Carson. 'I've had a great life.'

Julie Elliott on her graduation day with her sons Mark (*left*) and Paul. 'Life can be bitter and sweet. It's the people around that you love that make it so beautiful.'

Julie meeting former United States President Bill Clinton.

Senior Sergeant Julie Elliott with James Vance (*centre*), FBI Media Communications Unit, and Queensland Commissioner of Police Bob Atkinson.

Ambulance partners, and husband and wife, Colin and Robyn Heterick on their wedding day. 'The best thing we have in this job is friendship.'

On 9 May 2006 Paramedic Graeme Jones (*right, transporting Todd Russell to hospital*) looked after Todd Russell and Brant Webb after the Beaconsfield 'Great Escape'. Graeme was involved in assessing and managing the miners' physical wellbeing as they waited to be brought to the surface. (Courtesy Newspix.)

Paramedic Graeme Jones attending to a patient on Tasmania's Western Tiers.

The wrecked helicopter and casualties at the Tasmanian Western Tiers rescue involving Paramedic Graeme Jones.

The Mansfield family (*left to right*): Debra, baby Rachael, Mark, Sarah and Matthew.

Some of the Tasmanian Ambulance family working on the fence at Mark and Debra Mansfield's home not long before Mark's death. Also there, Paramedic Graeme Jones (*far left, holding orange mug*).

MARK MANSFIELD

'Life's Journey'

'He put others in front of him, not just in Ambulance, but everything he did.'

Debra Mansfield

Mark Mansfield was stunned. Over the years his job had taught him that nothing in life could be unexpected. But today, 29 August 1994, was vastly different from anything he'd ever encountered before. Shit! What would his friends say? His family? And Debra? God, Debra! She was just about to give birth to their first child. All the joy she was experiencing. All the excitement. All the wonder that lay ahead for the baby. What a mess. Why? Why now? Mark reached for his phone, dialled a number and soon heard the familiar voice of his mate and colleague Phil Brumby at the other end of the line.

'How are you going, Mark?'

'I've got rust.'

'What are you talking about?'

'Cancer. I've got cancer.'

***

Mark Francis Mansfield was born on 14 December 1954. He was raised on a 100-acre farm in southern Tasmania. Although the property was mostly thick bushland, there was still space for a few cattle, a dilapidated orchard and an adventurous boy. From the time Mark was big enough to stand up and walk in his favourite footwear, gumboots, he spent much of his time outdoors, relishing the freedom that country life offered. As he grew older, there were yabbies to catch, a go-kart to spin through the dust, even a boat to row on the Huon River, which lapped at the farm's feet.

Mark had two older sisters, Christine and Lyn. His father Doug, an engineer, had served in the Royal Australian Air Force during World War II, and his mother Doreen, had driven trucks in Brisbane for the Australian Women's Army Service (AWAS) that Doug delighted in calling 'Always willing after sunset!' Mark inherited both this sense of mischief and the feeling of belonging in a uniform. At times the two traits collided, like on the day that Mark and a friend connived during a Scout camp to lug a portable toilet onto the main road near where they were staying. The teenagers queued outside and pretended to hold their bladders, much to the bemusement and amusement of passing drivers. During the same camp Mark and his mate dressed in casual clothes and sneaked off to a pub for a sly beer. Believing they'd fooled all, they were deflated when a patron told them: 'I used to be in the Scouts too you know!'

Despite his mischievous streak, Mark was a dedicated student. His enjoyment of being a Scout drove him to achieve the movement's highest possible accolade, the Queen's Scout Award, which acknowledged success in personal challenges, and involvement in community projects. It was here that Mark first became involved in volunteer ambulance work.

Somewhat due to his father's wish that he pursue a 'solid profession', Mark undertook a three-year engineering cadetship after leaving school. While doing so, he continued his volunteer work, and such was his enthusiasm and interest that he eventually turned his back on engineering and joined the ambulance service as a full-time officer, stationed in Hobart. He was just twenty years old.

Within two months, Mark entered ambulance folklore when he had a fleeting meeting with a married woman in the middle of the Derwent River. Myrtle Walker already had five children, so when the pains told her a sixth child was on the move she needed to act quickly. By the time she was in the ambulance with Mark she knew that her impatient baby wasn't in the mood to be delivered in hospital. However, there was a complication. Just a month earlier the only road connection between Hobart's eastern and western shores, the Tasman Bridge, had partially collapsed after it was hit by the bulk carrier *Lake Illawarra*. Five people were killed. Hobart was split in two, which added pressure to the ambulance service because it didn't have a station on the eastern side. Although one was hastily set up, all hospitals in the city were on the western side of town, meaning patients in the east had to be transported by ambulance on special emergency ferries. It was on one such ferry that Mark said 'Push!' Myrtle obliged. In the wake of the bridge tragedy, the unexpected arrival of a new life on the Derwent was a timely story. Myrtle and her husband were so grateful of the role played by one particular ambulance officer that they named their newborn son Mark Francis.

Working on Hobart's eastern shore was exhausting. Even when off-duty, officers were on call and, consequently, there were times when Mark virtually worked 24 hours straight. Yet, he loved it. Ambulance was his life. His passion and willingness to stand up for what he believed in occasionally annoyed some of his senior officers, but no one could deny his enthusiasm and determination to be the very best officer he could be. His uniform was invariably immaculate, and his gentle, friendly manner endeared him to patients and colleagues alike.

There were the inevitable moments that tested his character. When operating by himself one evening, he attended a motorcycle accident. Being the sole carer he had no option but to recruit a bystander to drive the ambulance to hospital while he worked on the victim. The patient died on the way, prompting Mark to torment himself over how he could have done better.

He was a perfectionist, a professional. However, the cheeky glint in his big hazel eyes occasionally corrupted his views. While the Tasman Bridge was down, emergency services workers in uniform were allowed to travel on the ferries for free. Renowned for being careful with his money, Mark had been known to don his work gear outside of hours to ride the Derwent's currents. And there was also at least one pizza-shop owner who was never any the wiser when he received phone calls from the off-duty Mark.

'Hello, it's Ambulance Officer Mark Mansfield here. Could I please order . . .'

Mark would then dress in uniform, leave home, and pick up a Pan Supreme at a considerable discount.

He was not shy in making the most of things. Phil Brumby recalls one particular conversation in which his mate boasted: 'I met this girl the other night at a car accident. She's a real good-looker. I got her details and I'm seeing her this Thursday night. She was only marginally distressed but she still had to have a ride to hospital.'

In the mid-1970s Mark moved to Melbourne and stepped into a revolutionary world that reflected his skills and dedication. The training of Mobile Intensive Care Ambulance officers (MICA) was just beginning in Australia. MICA officers were considered the 'demi-gods' of the service because they were trained in coronary care procedures that until then had only been available in hospitals. Extending such life-saving techniques to on-road situations would revolutionise ambulance operations across Australia. It was a period of research, experimentation, intense learning and immense fun. Mark, one of the first officers to move interstate with transferable qualifications, was a pioneer during one of the most important times in Australian ambulance history.

After obtaining his MICA qualifications he worked out of some of Melbourne's busiest hospitals, including the Western General, which at one stage had three Tasmanian officers in the one unit. Administrators raised eyebrows the day they saw that one of their ambulances no longer had Victorian logos, but was decorated with ones representing the Tasmanian Ambulance Service.

No matter what the state allegiances, Mark's main loyalties lay with the people he treated. On one occasion, while taking a very sick child to hospital he was forced to stop at a heavy traffic jam along St Kilda Road. With time being wasted, he weighed up his options, and swung his vehicle in a new direction along a busy footpath. He was later reprimanded for his actions, but it mattered little to Mark, whose primary consideration was the welfare of the child.

By that stage, he had fallen in love with Christine, a nurse whom he'd met while transporting another patient. A week after their introductory conversation in the back of an ambulance, they went to a ball together, and within twelve months they'd quick-stepped into marriage.

After four years in Melbourne, Mark sought new challenges. He spent the next five years working in country Victoria where he and Christine, so familiar with holding other people's babies, cradled their own after the births of Sarah and two years later, Matthew. The celebration of life and the joys of seeing children grow often contrasted to the realities of ambulance work, in which Mark periodically dealt with death and trauma. However, even in the face of tragedies, smiles sometimes unexpectedly shone. When working in Sale, Officer Mansfield was elevated to local hero after he persuaded a charity to purchase a defibrillator machine for the area's ambulance service. On the very day that the equipment was installed, Mark used it to save the life of a man who'd suffered a cardiac arrest. In the years to follow, he would see defibrillators save other patients, but he could never have known just how close to him one of those people would be.

After nearly a decade on the mainland, Mark had returned to Tasmania with his young family. They initially settled in the northern coastal town of Burnie, where Mark was appointed Ambulance Duty Officer. He revealed himself as a true people person whose devotion to his job couldn't be questioned. Phil Brumby believed that when 'Mark joined the ambulance he found his reason to exist', while Mark's sister Lyn suggested her brother 'was always a boy with a toy in the ambulance service. He never tired of it and we never heard him complain about it.'

Ironically, his appreciation of the ambulance profession entered a deeper dimension after a case that he wasn't even involved in. Mark was at work when a visitor to his home collapsed. Within minutes two officers walked into the house with a defibrillator. They prepared the machinery, and placed electrodes on the patient's chest. Moments later the body was jolted by an electric charge. The patient responded. When Mark found out what had happened he took a very deep breath. Of all the cardiac arrests he'd either attended or been told about, he'd never imagined that one would involve his own father.

That incident remained firmly in his mind as he progressed through various roles in the ambulance service. However, while his career prospered, his personal life suffered. This culminated in the breakdown of his marriage while he was working at Devonport. It was a dreadful time as Mark left his wife and children to begin again in Launceston. In the ensuing years he worked on the road and as a supervisor in a number of centres throughout the state. He was also a superintendent in Hobart for a brief time.

His work eventually introduced him to another relationship. He was at the World Rowing Championships at Lake Barrington when he met Debra, a kind woman renowned for giving her time to others. She was on duty with the Army Reserves. What began with a campfire chat after a day's competition developed into a lasting relationship and two years later Debra and Mark married on Mark's 38th birthday. By then the couple had bought an old weatherboard house in Launceston that needed renovation. Mark thrived on the do-it-yourself challenge. From cornices to carpets, floorboards to fittings, backaches to blisters, he launched himself into the seemingly never-ending project. One day he would surround his castle with a picket fence. One day.

As he reacquainted himself with life as a husband, Mark became more involved in administrative issues, which eventually led to his appointment as the President of the Tasmanian branch of the Institute of Ambulance Officers. In this role he embarked on a special project that he hoped would bring the public and the ambulance service closer together. Unlike those at the front line of other medical services, ambulance officers rarely had the chance to formally meet the people they assisted. The very nature of their job meant they were the silent service. They would save, they would deliver, and then they would depart, frequently without the patient ever knowing their names. With the actions of two men large in his thoughts: Chris Tueon and David Galloway, the officers who saved Doug Mansfield, Mark wanted to change this.

By this stage Mark had fallen ill and, despite his determination to see his project through, his health demanded that he change his focus. He was 39 years old. All his experiences had led him to this point, but nothing could prepare him for what was to follow.

And it would all happen so very quickly.

***

The day after he was told that he had cancer, Mark was at Debra's side as Rachael Debra Enid Mansfield was born at Launceston General Hospital. A week later, just one day after his wife and new daughter had returned home, Mark went into St Vincent's Hospital to have a malignant section of his bowel removed. Another week later, he was told the cancer had spread to his lymph nodes.

As baby Rachael was learning to live, Mark was learning about life. Cancer was the biggest challenge he'd ever faced. But he was determined to beat it. Certainly, chemotherapy was something he hadn't had reason to think about a few short weeks ago, but now it simply had to be accepted; it was just another stage he had to go through before he could return to his usual busy self.

When he did regain full health, there'd be so much to do. He had a beautiful new daughter to raise, and two other children who were racing towards adolescence. He couldn't see them as much as he liked because they lived with their mother, but when the trio did get together, the reunions were wrapped in jokes and laughter. Sarah and Matthew considered their Dad 'a big ball of fun and warmth'. But he was also a teacher, often switching topics randomly to tell his children how they should treat a snake-bite victim, splint a broken leg, or bandage a deep cut. Mark loved seeing those looks of wonderment in his children's eyes when he explained something new to them. When they were younger he used to take them out on calls with him, and if the cases weren't distressing, he'd allow them out of the vehicle to carry his bags. This had been known to raise a few chuckles from some patients who'd joke: 'Oh no don't tell me you're that desperate for ambos!'

And then there'd been the times when Mark arrived home under lights and siren, much to the glee of his children. Despite all he had experienced in his life, nothing would ever give him the same satisfaction as fatherhood. It was a position of privilege, pleasure, and above all, much, much love.

Debra relished every moment that she had with her husband. Much of their time was spent in their own weatherboard home, which was beginning to lift its shoulders and stand up straight, but there was still a lot of work to be done, including the construction of the picket fence they both wanted.

However, Mark's illness was not diminishing. Three months of chemotherapy hadn't worked. Radiotherapy was tried. Mark started making jokes to Sarah and Matthew about the equipment that was being used. He remained positive. His eyes still had the glint of wickedness, and his smile curled with reassurance beneath his well-kept beard. He wouldn't be beaten. Simple as that. Or that was the impression he gave everyone around him.

But inside, he was having doubts. He asked his doctors questions. He asked himself questions, both as an ambulance officer, and as a patient. Whatever the answers, there was one irrefutable fact . . . he wasn't getting any better. Perhaps it was time to start making plans.

After he announced that he was taking Debra and Rachael on a trip to the mainland, no one in his family knew how ill he was becoming. He borrowed his parents' caravan, hitched it to the back of his red Commodore station wagon, and travelled through Queensland, New South Wales, Victoria and South Australia. Debra presumed the trip was partly to introduce her and Rachael to Mark's wide circle of friends and relatives, but in the midst of hugs, handshakes and hellos, one man, unbeknownst to everyone around him, was saying his goodbyes.

In the final stages of the two-month journey, Mark started feeling pain in his back and in a rare moment of openness about his illness, he confessed to Debra that 'something just isn't right'.

They returned to Launceston where Mark underwent a scan. The results exposed what he had quietly presumed: his body was riddled with cancer.

He tried to hide his disappointment. Nine months had passed since Mark had first been diagnosed with his illness. During that time he remained positive. He convinced others he would recover. He convinced himself. But now? What would happen to him? How long did he have? All questions, no answers. If time was indeed running out, he couldn't afford to be dragged down by misery. He had to keep busy, had to protect his family from grief. He arranged his own funeral, sorted out his will and finances, and while doing so, he thought of a Hollywood movie star.

Michael Keaton was perhaps best-known for his role as Batman in two box-office blockbusters, but Mark identified with the American actor for his portrayal of Bob Jones, in the film, *My Life*. Bob was a loving husband whose wife was expecting their first child when Bob discovered he had lung cancer and would die within a few months. While battling his anguish, he found comfort by recording his memories and wisdoms on videotape. This was to be a gift to his child.

Stirred by the poignant parallels to his own life, Mark approached his brother-in-law, Donald Adams, who had suffered his own tragedy when his first wife had died of cancer. Both men agreed a video would be a rich legacy, and so Mark began telling his life story in front of the lens.

The sessions usually took place in the living room of the Adams home. Mark, Donald, a camera, a tripod, and memories. Occasionally other members of the Mansfield family would join in. Doug came to listen, and Lyn and Christine took turns sitting with their brother as they remembered their childhoods. Amid the recollections, Mark also talked about the future, including the execution of his will. He had been meticulous in its drafting and wanted all his immediate family to benefit financially. He spoke too of his selfless wish for Debra to start again.

'If you meet someone nice, please don't spend your life thinking of me. Go ahead and marry him.'

Hours upon hours of tape were recorded. The sessions were draining but Mark carried on. Although he openly talked about death, he distanced himself by pretending he was someone else.

He told Lyn, 'Remember you're not talking to me, you're talking to the person next to me.' He rarely referred to himself in the first person, preferring instead to use 'we'. It was saddening for those around him, but it was the best way he knew how to cope. There were just a few times when he shook his head, or walked away from the camera saying: 'Sorry, I can't go any further.'

He gained strength from his own experiences as an ambulance officer. He'd seen people plucked away from life without the chance to leave anything behind. Mark was different. He thought he was lucky.

As the memories continued rolling, Mark's colleagues gathered to consider how they could best help their mate. They soon set to work, and a magnificent picket fence with arches and curves sprouted from the ground around the Mansfield home. Like the posts themselves, ambulance officers knew how to stand tall together.

By late July 1995, Mark was very weak. Yet he wouldn't slow down; there were too many things still to do. He went to a local jeweller and bought a ring for each of his sisters; he asked Debra to cut his hair, as she had for years, and dye the greying fringes back to their once natural dark brown; he went to a photographer and arranged a family shoot; he had a weekend away at Cradle Mountain with some mates; he visited other friends; he hugged, he laughed, he joked. He did all he could. But he didn't count the days.

By Friday 11 August, he had finally finished his work in front of the camera. He was gaunt, the whites of his eyes were yellow, and his skin was heavily jaundiced. He went for his regular trip to the doctor, and was away longer than usual. Instead of driving, he walked home alone, feeling the bite of a late winter's chill on his face.

The next day, he was admitted to Ward 3D at Launceston General Hospital. It was only now that Debra acknowledged that her husband was going to die. Until then, she had held on to that wisp of belief that somehow, somehow, somehow, a miracle would happen. But the awful reality tightened around her when Mark quietly insisted, 'If the doctors tell you they can prolong me, or they can let me go, I'd rather be let go. When my quality of life is no good, I want to be let go.'

A day later, Sarah and Matthew arrived. They'd been driven from Devonport by their mother. Under access restrictions they usually caught a bus to visit their father every second weekend. This time was unscheduled, and the children sensed the worst. Sarah was thirteen years old, Matthew eleven. This wasn't meant to happen at their age. They sat at Dad's bedside, and cried. Sarah, with voice trembling, read a poem that she'd written:

In the quiet times I think of you,
The fun we had, the things we do
Those things we'd share will always be
More than memories to me.
The memories of your loving smile
Would always linger for a while
So quietly with love
Your memory I'll treasure
Not today but forever.

Hours passed. Neither child wanted to leave. When Donald Adams walked into Mark's room, the sight was overwhelming. There were two red-eyed children, a very sick man smiling bravely, and a sea of tissues on the floor.

The following days were filled with more family visits. By now Mark was in agony. Through her tears, Debra saw her husband 'as this lovely looking fellow who'd been twisted and turned into a shell'. Ever since he'd returned from his mainland trip, he'd barely been able to lift Rachael. Now, his distress was much greater. Whenever his baby daughter was taken into his room, she screamed; perhaps even an eleven-month-old already knew too much. Debra made the heart-breaking decision not to take her little girl in to visit Daddy anymore.

Mark knew his time had nearly come. As he continued saying farewells to his family, his other family was also deep in his thoughts. He received a letter from the Director of The Tasmanian Ambulance Service, Grant Lennox.

'. . . you are regarded for your admirable personal qualities, your honesty and good humour and your infectious approach to life in general and all that has confronted you in your career and in your personal life. The remarkable and awesome courage and good spirit with which you have confronted your terminal illness has been a source of inspiration to all who know you and will never be forgotten.'

In the previous few months he had worked closely with Lennox to finish the project he'd begun with the Institute of Ambulance Officers. Now, all was nearing completion. A red heart-shaped pin with a gold lightning bolt had been struck. It would be presented by ambulance officers to the patients whose lives they'd saved by defibrillation. As Mark entered his final days, Lennox told him a name had also been decided for the official memento, 'We're going to call it the "Mansfield Pin".'

Mark was humbled. The name ensured he would never leave the Tasmanian Ambulance Service. He would be close to the hearts of both those who knew him, and complete strangers. That was the ambulance way.

On the morning of Wednesday 17 August 1995, Mark Francis Mansfield was undressed then wrapped in a warm, scented sheet. He was very, very weak. As Debra, his sister Christine, and two nurses softly massaged him, he died. He was just forty years old.

He would never be forgotten.

Several months after his son's death, Doug became the first recipient of the Mansfield Pin when he was reunited with ambulance officers Chris Tueon and David Galloway. Tears flowed in a room full of family members, administrators and journalists as they heard the story of the man behind the pin. It was said that day: 'Mark had left his mark.' Since then, the presentation of Mansfield Pins has become an annual event for the Tasmanian Ambulance Service. The occasions always attract media attention, and not only highlight emotive human interest stories, but educate the public about cardiac arrests and revival techniques.

Now, more than a decade after his death, Mark's influence still spreads far beyond the Mansfield Pin. Recollections of a father, a son, a brother, a husband, an ambo, a mischief-maker, and a lovable rogue still fill some Tasmanian homes with tears, and laughter. Family reunions, both of the blood and ambulance kind, prompt stories to be told and re-told:

'Can you remember when Mark was a little boy? Remember he used to dress up in an Indian uniform and pitch a teepee out the back?'

'Can you remember how Mark hated the sight of blood when he was young? And he became an ambulance officer!'

'And remember the time he faxed a copy of a twenty dollar note to a colleague who'd loaned him some money?'

Good times. Great times. Times to be cherished.

Reminders not only come in thoughts, newspaper clippings, photos, medals, pins, and boxes covered in dust. When Matthew smiles, he has a sparkle in his eyes that is older than his years. It sweeps his grandmother back to another age. Doreen Mansfield acknowledges, 'There are days when I've seen Matthew walk into a room and my heart would stop because he looks so much like his father.'

Both Matthew and Sarah have inherited another family trait as well. In 2005 Matt graduated with honours in aeronautical engineering at the Australian Defence Force Academy in Canberra. A career in the Royal Australian Air Force awaits him, while Sarah is a police officer. Two Mansfields, two more uniforms.

If her childhood dreams are realised, Rachael too will follow the tradition. She says 'When I grow up, I want to do something that helps the community. Probably an ambulance officer like my Dad. I know I never really knew him, but I think Dad was kind, nice and friendly. He looked like a very happy man. I think he was.'

Rachael is yet to see any of the videos that her father made. Sarah has watched snippets, and it was only towards the end of 2004 that Matthew sat down and hit the play button. On the screen in front of him was 'a different person from the one I knew as a ten-year-old. It gave me a different perspective. I actually laughed most of the time. I listened to some of the things he said and thought "Gee you were a cheeky bugger!"'

The contents of those tapes are for the family alone.

As Matthew and Sarah march through careers in uniform, Rachael, for the time being is content to sing and read her way towards high school. She also has a 'sister' to keep her company. A few years after Mark's death Debra applied to become a foster parent through a state government program, and after being mother to some 'short-term children' she accepted Kaylah, just two years younger than Rachael. The three have been a family for more than four years, with other foster children occasionally swelling the numbers. Motherhood, and working as many as three jobs at a time, ensures Debra a busy life. After Mark died she questioned how 'God could let this happen' but she soon decided there was no place for bitterness in a world that can take as quickly as it gives. Being from a large family she had long learnt to appreciate all that she has; her time with Mark reinforced this.

It can be easy to find faults in any person. Mark Francis Mansfield was no exception. Yet, he had a character that left an impression on others. His courage in the months leading up to his death will always be remembered by the Mansfield and ambulance families, and others who were touched by him during this time. He bravely showed that he knew how to die; to those who knew him best, that strength came from knowing how to live.

DAVE CUSKELLY

'From Choirboy to Coppa'

> 'There is nothing extraordinary about the things I'm telling you. The young men and women on the job are doing this every day.'
>
> Senior Sergeant Dave Cuskelly, Queensland Police Service

When Dave Cuskelly was ten, he was a member of the prestigious St Stephen's Cathedral Choir in Brisbane. His mother was so proud of him. His voice hovered with the angels twice a week, including one glorious evening when he sang alongside the world-famous Vienna Boys Choir. Some locals reckoned the Brisbane boys were better that night, but it's possible the odd wavering note was blocked out by Queensland parochialism.

Nine years later, the former choirboy was now a coppa. Though built like a rugby union breakaway: medium height, strong shoulders and chest, thick forearms, as he walked into the 'Big O' hotel in Cairns with his Senior Sergeant, he still appeared more a sweet-faced child than a man who could handle his own in a bar that was packed with suntanned trawler deckhands. These 'deckies' were hard drinkers. Their sessions began with a friendly round or two, and often finished with a wild swing or three.

It was Friday night, about 10 o'clock when Constable Cuskelly, on general duties, walked and sidestepped gently across the dance floor. The air was thick with cigarette smoke and more than likely the wafts from a few sticks of 'hooter'. A band was banging out heavy decibels. Dave noticed the lead singer, who through the haze and between the swaying heads appeared a 'good looker'. As he continued walking, Dave didn't realise that his Senior Sergeant had turned around and was leaving, assuming his young charge was at his heels. But Dave was heading the other way, deeper into the throng. After a few more steps, he saw a huge block of brawn leaning against the bar. Their eyes met, and despite his inexperience, Dave read the situation. He was in trouble! This block, a powerful deckhand, put his drink down, and headed towards his target. Five steps, four steps, three, two, one . . . Whack! He barged into Dave, pretending it was an accident.

'What did you do that for?' he said by way of an introduction.

Dave looked at the deckie, then glanced at the sea of accusing stares surrounding him. He considered his only two options. He could try conciliation, but saying 'sorry mate' certainly wasn't going to satisfy the deckie who was keen for some greater weekend recreation than slotting an eight-ball in the corner pocket. Dave's other option was 'to launch and see what happened'. Either way he knew he would be 'boxing on'. So he launched and grabbed the deckie with both hands.

'You're pinched, mate, you're coming with me.'

And then there was no need for any more chatter.

The Senior Sergeant heard the commotion and calmly looked over his shoulder. *What!* He'd expected to see his constable close behind him, not ducking and thumping, pulling and weaving in a ruck of fists. *Oh no, here we go!* He pushed through the crowd until he too was fighting with a Friday that didn't want to leave quietly. This was no longer a one-on-one altercation, but a good old-fashioned all-in brawl. As the minutes bled away, the fighters surged out onto the street. Back-up police arrived: one, two, three, four cars and more. Their windows were smashed, so too were faces. And somewhere in the midst of the fury, a young constable's hat was lost.

'I never did find it either,' recalls Dave. 'We ended up pinching about five or six blokes. Those sorts of nights could just happen. It was at the time of the Fitzgerald Inquiry (a Queensland State Government inquiry into alleged police corruption), and there was an anti-police feeling in some places. There were a lot of angry individuals in those days, and all I can remember is getting screamed at by them all.'

Dave was sworn into the Queensland Police Service on 4 July 1986—Cuskelly Independence Day. Until then, Dave, the youngest of four children, had lived under the roofs of Brisbane institutions. First, there was family, then the Boarding House at Marist College Ashgrove, and finally the Police Academy. When he arrived in Cairns for his debut posting, he was suddenly walking to his own beat. It was an age of discovery: how to drink grog, how to chat up women, how to cook, how to clean, how to budget . . . how to be his own man. But nevertheless, he still had to conform more than many others his age because his beat was also the one of a police officer. Once he celebrated his independence, Constable Dave Cuskelly was forced to grow up really quickly in a place that required 'frontier policing'.

Cairns was the end of the line for those travelling north. It had a constant flow of transients and others who set anchor without any real purpose. At the time there were 'a lot of people who had little to do and little to lose'. As a result, recreational illegalities such as drink, drug, and driving offences were frequent. Although idleness was a disease that struck some in the general population, it was never an issue for the police, who were always kept busy.

On his very first shift, an overnight shift, Dave saw the police station as a 'madhouse that I had no idea what I was doing in'. Whichever way he turned he seemed to find people being escorted, if not wrestled through mazes of furniture many of which were dotted with officers who were conducting interviews with offenders. Before the night had finished, Dave was also sitting at a desk, pounding away on a typewriter as a colleague questioned a person who'd been charged with wilful damage.

'Then this happened' *Tap, tap, tap, tap*. 'Then I went here.' *Tap, tap, tap, tap*. 'Then my friend said . . .' *Tap, tap, tap*.

Watching keys punch against a flaccid ink ribbon was an easy introduction compared with the keys he had to look after when he was on watch-house duty. At times he was the sole officer in charge of 65 prisoners in two cells at the back of the station. Arguments and fights were common. To finish these shifts without incident was akin to going into the 'Big O' in uniform without turning heads.

His early police education alerted his senses and assaulted his emotions. From a relatively sheltered Catholic background he was thrust into an environment that forced him to challenge himself in every possible way. He was a young man confronting, caring for, protecting, and observing the very best and worst of human behaviour. He was just nineteen.

Dave Cuskelly's classroom extended beyond the mainland to the Torres Strait Islands that are scattered like stepping stones between the northern-most tip of Queensland and Papua New Guinea. Dave spent three months based on Thursday Island. The mixing of Melanesian cultures and the logistics and time needed to move from one area to another revealed an entirely different side of policing from the one Dave had been raised on in Cairns. Nevertheless, he found some common themes, especially the possible outcomes when heat, alcohol and testosterone were combined. Thankfully he returned from his sojourn with both head and hat intact.

For an eager young coppa, nothing was better than the moments that were charged with adrenaline. Many of these came at night, which was, and still is the 'action shift'. Whenever Dave had to reverse his body clock, fighting off sleep was never a problem, as he and his colleagues were always busy, often frantic. In what was a dangerous period to be on the roads, drink-driving offences were rife, and it was rare for police to go through a night without at least one high-speed pursuit. One evening, Dave and his partner indicated to the driver of a V-12 Jaguar to pull over, but this only prompted the man to put his foot flat to the floor. The chase was on.

With seemingly no concern for the three passengers in his vehicle, the driver thundered through the streets, swinging his vehicle around corners to the roar of ear-shattering backfires as the car struggled with the sudden deceleration and acceleration. Surely few residents in the area could have stayed asleep amid such a din.

The chase eventually ended in the front yard of a house. In the aftermath, Dave and his partner were blown into the media spotlight when a local radio station took a call from a listener who protested: 'I know the police have a difficult job, but I think shooting at speeding cars is going a bit far.' Dave was promptly called by his superiors for an explanation. His gun and ammunition were inspected, but there was certainly no room for a V-12 Jag in his holster!

In some other chases, horsepower made way for the good old-fashioned foot slog. Like the day Dave was off-duty and eating a burger at Hungry Jack's with his girlfriend, who would later become his wife. He watched with interest as two of his colleagues pulled up a vehicle. A man got out in a hurry, threw a large parcel of dope up in the air, and sprinted away. Dave immediately dropped his food and gave chase until the assailant was caught. These were the episodes that Dave relished: the action, the arrest, the thrill, and the sense of satisfaction.

But the job wasn't always ripped from the pages of the *Boy's Own Annual*. Seven months into his career, Dave attended a bus crash in which eight high school students were killed. His role was to assist with identification of the dead, and to ferry the dead and injured to the Cairns Base Hospital. If this wasn't distressing enough, the sight of a mother identifying her child most surely was. Dave stood there thinking, 'These kids are only four years younger than me.' The link with people he'd never known was suddenly very personal.

Every one of his experiences during his first year was an important part of his climb up a steep learning curve.

'In that first year I went from zero to 100. You think about it. I was nineteen years old, but I was going to domestic disturbances and being asked my advice. I didn't know. I was nineteen! By the end of that year I'd been to murders, road fatalities, some pretty ordinary jobs. It gave me a slant on reality that has stayed with me. Even now I have very little time for people who introduce drama into things that don't need it. I say that to my kids: "There's enough drama in this world without you introducing any more."'

Dave has two children, Amelia and Daniel, both of whom are approaching adolescence. They are the products of a relationship that began in a nightclub during Dave's eventful first year in Cairns. When the off-duty constable met Linda at the 'Playpen' it seemed brushes of shoulders didn't always lead to confrontations. They have now been married for seventeen years, which Dave acknowledges 'is a bit of a record in the Police Service'.

It's inevitable that some police families will be exposed to behaviour that the general public never experiences. When Dave was the Officer in Charge at Port Douglas, a few years ago, he and his family lived in a compound consisting of a house and a station. Linda was paid to clean the work area, and in one week alone she swept away the glass from three shattered police car windows that had been kicked or punched out by offenders. The eighteen months of compound life gave her a thorough understanding of her husband's job. For young Daniel, the comprehension wasn't as strong, but the three-year-old certainly knew right from wrong. One day when he was sitting on the steps of the station, he watched his father escort in a prisoner.

'Dad, is he a baddie?' he asked.

'Yes mate, he's a baddie.'

Dave felt comfortable with such exchanges when they were in the appropriate domain, but his views changed abruptly when his family's space was intruded on.

'There were times where I'd be pushing the kids in a trolley through Coles or we'd be in public somewhere and we'd see someone I'd boxed on with, or pinched. Sometimes they'd say something smart or abusive, or they'd just stare at us. That used to frustrate me. That's why it's hard for police to go to smaller places for more than a couple of years. People become too familiar with you and what you're about.'

After his first year in Cairns, Dave worked in Brisbane during World Expo 1988, then he returned to Cairns. By the time he was 26, he was a sergeant, a rare rank for someone so young, and an example of the merit-based promotion system that had only just been introduced. He spent the entire 1990s in Queensland's far north before he came back to Brisbane, where he is today. During those years, he rode the highs and lows of being an officer at the front line and, in doing so, he became a student of life, learning about himself and others 'in a way that you could never be taught at university'. He'd been spat at and kicked, punched and 'split open'; he'd been called a prick, a wanker, and other words unprintable. He'd watched helplessly from the back of a plane as a prisoner made a mad lunge at an emergency latch; he'd dived into a crocodile-infested inlet and chased a car thief; he'd stopped an Aboriginal in custody from hanging himself; he'd pulled his 357 Magnum on a man wielding a knife; he'd searched for an elderly woman who'd vanished off a reef; he'd gone riding on horseback looking for marijuana crops; he'd protected the Olympic flame; he'd been flown into remote communities to quell riots; he'd informed a wife that her husband had been murdered while her children prepared for a slumber party; he . . .

He was simply doing his job.

The memories of some encounters have frayed over time, but others remain clear. There are the recollections of high farce, such as the time that Dave boarded a coast guard vessel to search for a man who allegedly had a concealable firearm on his boat. The rest of the story best comes in Cuskelly coppa slang:

'This coast guard boat was a huge, clapped-out piece that would only go eight knots. It was at night. We were creeping up on this bloke, and we were supposed to hit the spotlight so that when we did the entry onto the boat the bloke would be blinded. So we got pretty close, and when it was time to hit the spottie, one of the crewmen, they are volunteers, hit the fog horn instead. We blew our element of surprise and the bloke was nowhere to be found on board.

'The next day, or the day after that, we were going up to do the same job and we had the doggie (dog handler and dog) on board. We were going to pluck this bloke when he goes past the other way in a dinghy with an eight-horsepower motor. We turned our boat around, and all of a sudden you've got this bloke in his little boat doing six knots flat-out going *nnnnnnnnnnnnnnnnn*, and we're wound flat out going eight knots *dddddddddddddd*! It's the slowest high-speed chase you'd ever see. "Water Rats" it wasn't.

'Anyhow, the bloke pulls up at the yacht club and does the runner. We jump off, then the doggie jumps off and goes straight through the trampoline of this catamaran and into the water. We ended up having to pay for the rip. Anyhow the doggie was embarrassed and his dog was fairly upset. It was hilarious, and as police we were not letting the opportunity pass to pay out on each other as it was happening. We ended up catching the bloke too!'

There was also the occasion, before the days of holding cells, when an offender who was being processed at Port Douglas Station decided to chance his luck and escape out the front door. Reacting quickly, Dave and his partner reached the offender in a couple of steps but, in no mood to be contained, the man started swinging, forcing Dave's partner to use capsicum spray. The offender screamed and, with eyes burning and rage growing, he sprinted across the road and tried leaping over a fence into some shrubs. However, he miscalculated his jump and was momentarily jack-knifed at the top with his legs on one side and his torso and arms on the other. Dave's partner reached him, and grabbed onto his shorts, the only piece of clothing he was wearing. The offender squirmed and wriggled his way free, but in leaving the police behind, he'd also surrendered his shorts. He ran through the shrubs, and after bursting past some palm trees, he discovered he was on the pool deck of a five-star resort crowded with tourists, who were stunned to see a naked man in tears, diving into the water. This certainly wasn't part of their package deal! Upon surfacing and reaching the pool's edge, the man was met by Dave and his partner, a pair of handcuffs, and . . . his shorts.

But these moments are rare when compared with those of grief and trauma that every police officer on the beat encounters. Dave has attended the goriest incidents without being affected, yet he 'feels crook when I watch my young fella get a needle'. It's the personal link that can stir his emotions, or intriguingly, the finest details. He explains:

'I went to a double murder suicide. I was the second or third person there, and what struck me was the smell of gunpowder in the air, and [one of] the deceased's cup of tea was still steaming. Police inherently roll up to jobs after the offence has been committed, but those details put me in the time frame. I also remember a triple fatality I went to. One of the deceased had a watch on with a second hand that was still moving. It brought reality to the situation. It made it definite, made it real. I don't know why I notice those things.'

Senior Sergeant Dave Cuskelly is now Officer in Charge at one of Queensland's busiest stations, Brisbane City. He is responsible for the activity of 170 uniformed officers within the division. Adventure has made way for management and, as a result, Dave rarely gets out on the road any more. However, his experiences when at the front line aren't just memories because they are used to help educate the current generation of young officers. The most important advice that Dave can give is, 'Always be positive, confront negativity, and never assume anything.'

He has been a police officer for twenty years, and could be forgiven if he would now rather a cosy job in an air-conditioned office with a view. But that would take him away from the one basic essential that keeps him in uniform . . . people. Despite colliding with the uglier sides of life, he is adamant that 'most people have a definite level of decency'. It's these people who motivate him to be the best police officer he can be.

Dave Cuskelly, once a boy with a pure church voice, has grown into a down-to-earth bloke who has a need to perform for the masses. No applause needed. Just a simple 'thank you' will do.

DANNY CARSON

'Memories'

'What does fire mean to me? It just means I've had a great life.'

Danny Carson, Chief Fire Investigator, Queensland Fire and Rescue Service

The immortal American golfer Bobby Jones, regarded by some as the greatest player of all time, once said that his sport 'is played on a five-inch course—the distance between your ears'. Danny Carson travelled thousands of kilometres across the world to discover this. When he was just nineteen Danny was lured away from the harsh redness of outback Queensland by the rolling green hills of Ireland. He went to the Emerald Isle to improve his golf but during eleven months abroad he didn't once swing a club. Instead, he discovered the not-so-delicate touches needed on a pint glass. Fun came before future, and with it came the knowledge that professional sport wasn't going to be Danny's career.

Danny is a matter-of-fact bloke who loves to laugh. He is married to Bronwyn, has two teenage daughters Angela and Olivia, and still unleashes the driver when he finds the time. He is also the Manager of Fire Investigation at the Queensland Fire and Rescue Service. His cheerful nature is a wonderful foil in a job that can often be full of darkness.

'Fire investigation is quite unique,' he says. 'Firefighters may go to a house fire where there's a fatality. After their job is done, they pack up their hoses and go, whereas investigators may be there for six or seven hours working around a body. Then we write about that body in our reports. Then we go to the coroner's court and talk about that body again. It's not a one-off thing for us; it continues on. It does affect blokes, but nowadays we have ways of working it out with special services. But 25 years ago you would have been called "a pansy" if you were seen not to be dealing with it, especially in Mount Isa, which is a pretty tough town.'

Mount Isa, nearly 1900 kilometres north-west of Brisbane, is one of the most famous mining towns in Australia. Silver, zinc, copper, lead and blue collars are all abundant. Despite it being an unwritten Isa law that any son will follow the light cast by his old man's helmet, Danny had no interest in 'going down a big hole and working like a badger all day and all night'.

As a boy, he was only drawn by the holes that he sized up on golf courses. By the time he was a teenager he'd won some tournaments in northern Queensland, and worked his handicap down to near scratch. At the time a vein of rich talent had been struck in Mount Isa. Just a few years older than Danny, a broad-shouldered blond youngster was beginning to make headlines. His name was Greg Norman. However, while Norman pursued a career walking the fairways, the weight of Isa tradition eventually pushed Danny along the path to the mines where he did a boilermaker's apprenticeship. After he finished, the golf tees still beckoned, so he headed to Ireland, where relatives had arranged for him to begin a traineeship with a course professional.

'I actually never got to meet him,' says Danny, laughing. 'I had all good intentions, but there were too many diversions. In the end I had no money, no golf clubs—I'd hocked [sold] them—so I came home. But I had a great time. To be honest I didn't have a dream about it. I didn't have a dream about anything. I was having too good a time to think too hard.'

Reality stared at him once he returned to Mount Isa, and whichever way he looked, he saw a future in the mines. He delayed the seemingly inevitable by drifting between odd jobs, including one as a hospital wardsman, but when he left the sterile corridors, dusty shafts were still awaiting him. Or so he thought. He now jokes about how he was saved from taking that passage. He was a beneficiary of the age-old 'it's not what you know, but who you know'. By this stage he was engaged to Bronwyn. When her boss, a plumber, suggested that Danny apply for a position with the local Fire Brigade it seemed as good an option as any other he'd considered. Until then, Danny hadn't ever thought about firefighting even though he'd been raised just one hundred metres from the Mount Isa Fire Station. Nevertheless, he replied to an advertisement in the local paper and waited to see what would happen. Within a week of sifting through the applications and interviewing candidates, the Fire Chief strolled down the road to the Carson home and offered Danny a job. All these years later one particular Fire Investigator admits that his entry into the service had some suspicious sparks.

'Bronwyn's boss was a good friend of the Fire Chief, so I have a fair idea a few strings were pulled. The day I started was the day I was riding a fire engine—24 April 1979. Those were the days when we still actually rode on the back of the truck. I had no training whatsoever. We did the occasional drills out the back of the station, but the old blokes wouldn't do it. They'd just say, "We've done all that. We don't need to do it anymore." They didn't seem to have a care in the world, but they were excellent at their job. All my experience came from what the old blokes did on the job. On each shift there'd be an officer and two firemen and one truck. If we needed more men, off-duty officers would be rung. They'd go to the station, get the second truck and come out and give us a hand. I was so nervous when I started. I did two day shifts, then I was on night shift for two nights, and I never slept a wink; every time a car drove past my eyes were like golf balls.'

Almost every day at Mount Isa offered Danny new experiences, including sliding down the steel pole that reached from the engine bay floor through a hole into the firefighters' living quarters on the first floor. This was the emergency route, the quickest way down for firefighters to be on the truck and ready to go. Nowadays only a few stations still use poles, but their standing in brigade folklore will never be forgotten. At Mount Isa, there was the legendary day a firefighter mistimed his grabbing of the pole and he plummeted head-first down the hole. Shocked from his drowsiness, he recovered quickly enough to slow his progress. Hanging betwixt and between he looked for support. For years afterwards, a pair of black boot-marks on the engine bay's ceiling were a reminder of his misfortune. Even when the station was painted, no brushstrokes were ever allowed near these symbolic treads.

Danny grins when he recalls such moments. As you have no doubt realised by now, humour plays a lead role in the character of many firefighters. It can help them overcome the horrendous parts of their jobs, and can lighten the mood during the long, sometimes frustrating periods of downtime. When Danny walked into Mount Isa station he entered a world of practical jokes and quips from tough men who never saw the need to waste words. None were more memorable than Lionel, whose trademark was a soggy piece of grass in the corner of his mouth that twisted in time with his chewing. When the Isa was once hit by a petrol strike, Lionel rode his horse to work, and tied it up at the side of the station. As the days without petrol passed, the hooves clipped on concrete, and a pile of manure grew. When he was asked by a senior officer to clean away his horse's mess, Lionel scattered it everywhere by cutting it with a lawnmower. This was the same bloke who, when instructed to paint the common area of the fire station, slapped a bright pink on the walls. His defence was simple, 'You didn't tell me what colour!' Lionel also used to fall asleep in night lectures given by the Chief Officer.

'Why have you come if you're not interested,' said the officer.

'Because this is the best therapy I can find for insomnia!'

While Lionel and many similar characters have disappeared from the fire service over the years, people such as Danny ensure a bygone era lives on through the stories that are told. Not all yarns are painted with mirth. Understandably, many of Danny's most vivid memories of his bush education are of accidents and tragedies in an area which posed its own special problems. On many occasions, answering a call-out wasn't just a matter of donning the work gear, and driving out in the truck under lights and siren. The sheer isolation of the area often meant that a job 'just down the road' was well beyond the corner store and into a vast red sea of emptiness. Danny remembers one particular job, a motor vehicle accident in which there were multiple fatalities, but some people were still alive. After receiving news of the accident in the late afternoon, Danny and his partner prepared all their hydraulic rescue equipment, and then flew with the Royal Flying Doctor Service (RFDS) to Dajarra, a settlement 120 kilometres away, but still half an hour's drive from the accident scene. With no room in the RFDS vehicle, they hitched a ride in the four-wheel drive of the local publican. After arriving, they worked for two hours in the desert's chill cutting people out of the crumpled vehicle. Job done? Not yet. They drove back with a policeman to Dajarra, where they were told by the Flying Doctor, 'There's no room on the plane. You'll have to find your own way home.'

Just as it seemed they'd be staying for a night in Dajarra, they were approached by two ambulance officers.

'We're going to Mount Isa, and we need some drivers because we've got some patients that need looking after. Can you do it?'

In such an isolated part of Australia, this wasn't an isolated incident, and nor were the moments that would perhaps shock people in other parts of Australia. When high speed and open roads combined with the unexpected, the consequences were invariably tragic. Danny explains, folding his arms across his chest:

'Sometimes it was hours before an accident was found, and then we'd spend more hours getting there. And when we did, it wasn't always people who were the worst off. The amount of bullocks, kangaroos and pigs that were hit made for some pretty ugly situations. Think about this. I've been to jobs where there are both horses and people stuck in vehicles. The horse is kicking, and the only way to get to the people is . . . well, you have to do some horrific things to knock that horse off before you can do anything for the people. The horse has to die, and you have to do it. Many people, especially in the cities, will never see or comprehend that side of the job.'

The sparkle in Danny's eyes has gone, but it soon returns as his memory steps back through the towns in which he worked. Mount Isa, Cairns, Maryborough, back to Isa, then after eleven years of bush service, he shook hands with Brisbane as a Station Officer at Kemp Place, the city's main station. It is not uncommon to walk through the rolling maul of city life and identify a country person. It may be his cracking leather boots, or the way he struggles to keep pace with the throng that surges from a street corner when the lights change, or maybe it's his wide eyes staring up and up and up at a skyscraper. When Danny Carson from the Isa rolled into Brisbane, he gave himself away as a bushie by counting.

'It was a shock to go to a station where all of a sudden there were 23 men starting one shift at one time. I just couldn't get a grasp of it, and was thinking: "Hang on, what do all these fellas do?" Once we got to know each other, I used to bait them a bit. I used to call them the "footpath firemen", and that would always get a reaction. They got that name while I was still in the country. Whenever we saw any metro firies on TV they were always holding hoses on the footpath. It wasn't until I got down there that I realised they had so many blokes that the others were on the inside!'

Footpaths, roads, and corridors eventually led Danny to his position today. In an era when night-time television overflows with hyped-up American crime investigation programs, Danny lives and works in an Australian reality. After arriving at a fire scene, he generally follows a standard procedure. He and his team inspect the outside of the affected building before they talk to owners, witnesses, and various other services that may be involved, such as Police Squads [Scientific, Crime Scene, and Arson]. From these interviews, they determine the external and internal appearance of the building before the blaze. This includes the positions of contents, such as furniture, machines, electrical goods. Diagrams and sometimes models may be made. Once inside the building, they begin investigations in the area where there is the least amount of damage and methodically work towards the areas most affected. They examine the direction of burn patterns. They sift, they observe, they smell, they sense, until they hopefully find both a starting point and a cause. A magazine underneath a faulty toaster? A cigarette in bed? An accelerant? Some causes are easy to determine, while others will never be known. It's painstaking work. It can take a matter of hours, or may stretch into weeks. Danny acknowledges:

'It can get frustrating when that one piece you're looking for just doesn't appear. We get a lot of people telling us how a fire happened, even if they weren't in the house or the building. People love to give you stories. In a lot of cases we ask the obvious question of the resident of a burnt house "Do you smoke?" It's amazing how many people will say "Yes, but not inside." Then you go in and find an ashtray next to the lounge!

'If we find there has been something untoward going on we'll immediately hand it to the police. It's sad to know there are people out there who want to be so destructive. The toughest part is going to the fatalities. It doesn't matter whether they are young or old. The loss of life at any age is sad. I don't do it anymore, but I used to wonder: "What was this person thinking when he knew he was trapped?" The bodies in house fires are terrible. You just hope that the death was quick and the person didn't suffer too much pain. Usually smoke kills them long before the flames.

'One of the main questions I'm asked is: "What are the most common causes of fire?" The answer is simple: It's men, women and children, whether it's deliberate or not. It can be complacency, foolishness, lack of knowledge. I hear people all the time saying, "I didn't think this could happen to me!" Well it happens every day of the week. It happens to anyone. No one should ever become complacent because if they do, fire can happen to them.'

It has now been a quarter of a century since Danny Carson became a firefighter. As with so many in the emergency services, his experiences have taught him to appreciate how precious and rich life is. To quote another famous American golfer, Ben Hogan: 'As you walk down the fairway of life you must smell the roses, for you only get to play one round'.

THE PITMANS

'Keeping the Faith'

'Mum would say to us: "You might be a police officer, you might be a church leader, you might be this, you might be that, but don't forget you're a Pitman, and I expect you to act like a Pitman."'

Superintendent Grant Pitman, Queensland Police Service

The public square near the Gare Matabiau in Toulouse was crammed with people, placards, and political platforms. Most in the 200-strong mob were university students. Two young clean-cut men walked past. Dressed in black shoes, white shirts with ties, and dark suits with name tags on their lapels, they were easy to notice among the mass of jeans, T-shirts, caps and beanies.

'Americans!' yelled some of the protestors.

What a pleasure it would be to hurl capitalists into the fountain on this day of statements.

The young men were jostled and grabbed. They were defenceless against, ten, twenty, thirty sets of hands pulling and pushing them through the cheering crowd.

One of the young men shouted.

'No, no. Not American. Je suis Australian. Take my camera, take my projector.'

But he continued being swept along the wave of rebellion. It seemed his was one voice that wouldn't be heard.

Twenty-six years later, Kendall Pitman smiles when he remembers that day in the south of France.

'We actually didn't end up in the fountain. Fortunately, an American missionary came running through the crowd belting people out of the way like ninepins. It helped that he was a wrestler and was a big strong fellow.'

A missionary who was also a wrestler? Perhaps it wasn't an altogether unusual combination considering, at the time, Kendall was a missionary himself. While he had chosen to follow the Lord, he'd also decided to walk in the steps of his older brothers, Grant and Garth. All three had gone to the Queensland Police Academy before taking two-year placements to deliver the messages of the Mormons. The trail didn't end there; younger brother Stacey would also follow the same route.

All four resumed their police careers once they'd finished their missions, and today they are in specialised positions in the Queensland Police Service. Superintendent Grant Pitman is with Communication Operations, Support Command; Superintendent Garth Pitman is Business Manager, Operational Policing Program, Ethical Standards Command; Inspector Kendall Pitman is with the Inspectorate and Evaluation Branch, Ethical Standards Command; and Inspector Stacey Pitman is Regional and Education and Training Officer of the North Coast region.

'There are parallels between our church and the police,' says Garth. 'In both, we are genuinely trying to help people. When I went on my mission, I didn't see it as conflicting with my police career, but as a continuation of it.'

The beginnings of this curious lineage can be traced back to the 1950s when Brian Pitman left his job as a grocery storeboy in Canungra, a quiet village that is now cradled in the hinterland of Queensland's Gold Coast. He headed to Brisbane, and after his National Service, he joined the police. Around this time he met a young woman, Patricia Saunders, at the Cloudland Ballroom. They soon married and had their honeymoon on board a train rattling to Cairns, where Brian had accepted his first police posting. He was a Methodist, his wife a Presbyterian, but when two Mormons came knocking on their door not long after they arrived in Cairns, they were so encouraged by what they heard that they shifted their faith to The Church of Jesus Christ of Latter-day Saints. In doing so, Brian and Patricia Pitman had laid a path that their family would happily tread in the years to come.

As Brian moved from post to post around the north Queensland area, Patricia performed forever increasing general duties as the mother of a burgeoning family. Brian had initially wanted six children, Patricia thought two was enough; so they compromised and had eight (six boys and two girls) over eighteen years.

Grant was the oldest. Within six years of his birth, Garth, Stacey and Kendall were all in line for his hand-me-downs. Although their father didn't talk much about his job, the boys were still exposed to a country cop's life. Garth recalls going out to an incident in which his father and a colleague trawled a river for bodies following a triple fatality motor vehicle accident. Such harsh exposure for the boys was rare, as they were more accustomed to playing with dad's radio and handcuffs, and running along the corridors of the Cairns, Bundaberg, Dimbulah and Munduberra stations. When just twelve, Grant revealed he had the keen sense of observation needed to become a police officer when he detected all wasn't quite as it should be outside the Munduberra Police Station. He promptly walked inside and told his father: 'There's a man lying down outside the front fence. He's got ants all over him. I think he's dead.'

'No, he's probably drunk and sleeping it off,' replied his father.

Grant accepted his Dad's assumption, but curiosity soon drove him to make a second inspection. Yes, the man was indeed dead. He'd suffered a heart attack.

Two years later, Brian accepted an administrative job at Police Headquarters, and moved his family to Brisbane. By then there were seven Pitman children (five boys and two girls), meaning there were no empty corners or shelves in the small four-bedroom housing commission home that they crowded into. However, Brian still managed to have his own study, albeit the lounge room, which was invariably full of quarrels, games, or a child searching under the furniture for a lost sock or shirt.

The Pitmans were a tight-knit family of high principles. Brian and Patricia instilled in their children that 'families are forever, and no other success can compensate for failure in the home'. Their religious beliefs reinforced this. Amid kicking footballs, reading Scout maps, squirming under Mum's clippers, licking cake bowls and all else that came with growing up, each Pitman child was taught that the Church was their second home.

The importance of education was also emphasised. Brian, who'd left school in Grade Five to work in the Canungra grocery store, recognised that the days of walking into jobs without qualifications were shortening.

When he was fourteen, Grant became a cadet at the brand new Queensland Police Academy. The system no longer exists but at the time Grant was the youngest ever admitted. It was 1972, a time when the State Government was introducing a more academic approach to policing. Cadets, who were paid a small wage, could complete Years 11 and 12 of their school studies at the Academy before deciding whether or not they would stay on to do a further year of police studies. After this, they were posted to stations in trainee roles until they reached their nineteenth birthdays, at which age they could be sworn in as police officers.

The Pitman name soon stretched broadly across the Academy rolls, as Garth, Kendall and Stacey all followed their oldest brother.

'It was a great way of life,' recalls Stacey. 'There was plenty of sport, and best of all we got paid! We could buy our own lollies and get our own haircuts as opposed to Mum giving us a crew cut. We could even buy our own underwear. Mum used to make our underwear, and I got a lot of hand-me-downs. There were a few holes in them by the time I got them!'

Things had changed when Garth took his first educational steps towards a police career and cadets could no longer do their senior schooling at the Academy. So Garth began each day 'spick and span in uniform' on morning parade at the Academy, then he hurried off to the local high school before returning to the Academy in the afternoon for more drills. Although the systems of learning were different for each of the Pitman boys, there were some common themes, none more than discipline. On his very first day at the Academy, Kendall was standing with his arms folded outside the 'Fish Bowl,' a large room surrounded by glass in which supervisors could do work and look out at the future of law and order. His posture immediately raised the ire of an uncompromising sergeant who called the new cadet to his side. Kendall was then forced to stand at attention while his superior shouted and snarled at him, drawing the notice of all would-be officers within earshot. It was the last time Kendall ever folded his arms at the Academy. On the same day, all rookie cadets were addressed by another instructor, who was as surprised by Kendall as Kendall was shocked by the instructor.

'Who here has never had a drink?' asked the instructor.

Kendall was one of just a few who raised his hand, and consequently, a few eyebrows.

'Well,' said the instructor, 'you have had a very sheltered life and, to be honest, I don't know whether you'll ever make it as a police officer!'

But Kendall, as with all his brothers, adhered to the teetotalling doctrine of his religion. They also put in long days to follow their faith, which sometimes meant starting as early as 6.30 in the morning with seminary classes or youth studies.

There was also another family practice, and although it wasn't a religion as such, Australia's culture ensured it a place of worship among some of the Pitmans' cadet colleagues. This was sport, most prominently the football codes, rugby league and rugby union. Although their beliefs prevented them from playing on Sundays, the Pitmans still had another six days to charge headlong into games. At one stage or another, various combinations of brothers played in the one Under-18 Academy team. Grant, Garth and Kendall were all hefty enough to be forwards, while Stacey, the slightest of the clan, was a winger.

The games against other teams from Brisbane could be rugged affairs, fuelled by the satisfaction that some players felt when taking pot shots at would-be cops. But in landing the odd underhanded thump, they exposed themselves to retaliation from a team bonded by both institution and brotherhood.

'We would go to a match and you'd hear the other team preparing for it,' Kendall recalls. 'They'd be screaming in their dressing-room: "Kill the pigs! Kill the pigs! Smash the pigs!" Our players would look at each other and you'd see this steely attitude. We always played hard and fair and it always seemed more of the opposition got carted off than we ever did. We had a strong motto: "One in, all in." If someone wanted to punch a cadet, we defended that cadet. One particular day an opponent punched me. What he didn't realise is that when he punched me he not only punched a cadet, but he also punched a Pitman. I can't remember how many other Pitmans were on the field at the time, but there was a huge fight. Although it was a tough way to show it, it revealed our family motto too: "Pitmans are one".'

When they each turned nineteen all the boys had to go their own way. Not only was this the age at which they could become police officers, but they were old enough to become missionaries. The selection process for the latter involved a number of stages: an interview with a church leader; submission of papers to determine suitability; more interviews, including ones with family; an assessment of finances to ensure funds wouldn't run short when on a mission; and finally the presentation of all material to leaders at the Church's centre in Salt Lake City, America, where a decision would be made. Once accepted, a missionary underwent training, which included a language course if the posting was to be in a non-English-speaking country.

Over a span of six years, Grant went to Melbourne, Garth to Canada, Kendall to France, and Stacey, Perth. The spiritual trend they set caused enough concern in the police corridors of power that prior to his departure Kendall was invited to see Commissioner Terry Lewis. He acknowledged that leaving the police was a 'heartfelt decision', but he felt certain that his mission work would make him a better person because of the number of people he'd meet. After all, not even the police knocked on as many doors as Mormons did.

The life of a Mormon missionary was hard work: rise early for personal study; more study with your companion; then out onto the streets knocking on doors for hours on end, hoping that someone would be interested in listening to you. When there was interest, a person was told about the principles of the Gospel of the Mormon Church, and if that interest grew into a belief, he could ask to be baptised by a missionary. Grant explains the process.

'We saw the mission work as a spoke in the wheel. It was just one part of helping us to become well-rounded. You could go two to three months without anyone wanting to talk to you. You'd go back to your apartment after ten hours of trying, but you'd never give up your resolve because you knew someone, somewhere, would want to talk to you. Some missionaries go home without ever making a baptism. It can be tough.'

The Pitmans had better fortunes, which are best highlighted by Kendall's baptism of an 80-year-old woman who 'felt the time was right'.

After completing their two-year missions, each of the brothers would return to Brisbane eager to fit back into regular Church life and resume their police careers. This posed some challenges. Grant was the first to reapply. After conducting an interview with him, a police panel recommended that their interviewee not be allowed to return to the service, but this was overruled by Commissioner Lewis. Two years later, Garth was asked by a similar panel: 'Will you work on Sundays?'

'Yes, if I'm asked I will work on Sundays,' replied Garth.

'But what about church?'

'Well, if I'm not working on Sundays I will go to church.'

In the same interview he was also asked if he would shoot someone.

'If I was in a position which demanded it, yes I would. If I was abiding by the law for the good of the community, then yes I would do it.'

By the time Kendall and Stacey reapplied there was a greater level of acceptance by the powers that be. However, Kendall had other small hurdles. He arrived back from his mission with just enough money to buy a return ticket from his home to Police Headquarters to attend an interview. Although he can't recall for certain, it's highly likely he wore the same threadbare shirt that nearly had an untimely wash in Toulouse. It was a relief when he started his re-education at the Academy only a month after returning to Australia as he could again earn an income.

The brothers all began their police careers as general duties officers, but as time passed, they were made increasingly aware of their father's belief that 'the future of policing will lie in the hands of those who are educated'. Leading by example, Brian attended night school in his thirties to complete Years 11 and 12. He then began a Bachelor of Arts degree at the University of Queensland, often leaving home before sunrise to start a shift and returning late at night after evening lectures. In between, he also fitted in his church and family commitments. Inevitably, all four police brothers followed their father's lead by undertaking tertiary studies. Grant was the first. At the time, violent political demonstrations were rife. Queensland Premier Joh Bjelke-Petersen was under fire, and so too were the police following ongoing accusations of corruption.

The protests frequently had an element of university students. Grant remembers it was a volatile time to be both a student and a police officer.

'The University of Queensland could be a hostile place. There was a common room where journalism students met. On a wall were the names and pictures of the "10 Most Wanted Police", and my father was one of them. He'd locked up his share of student protestors at the time. There were some students and lecturers who just didn't want to accept us. I regularly went straight to uni from work. I used to throw a T-shirt on over my uniform, but you could still pick the police pants and the police boots. I was in a lecture theatre one day when this fellow came in and said to the class: "You have three police in here doing a law subject. We want you to throw them out. They shouldn't have anything to do with us."'

Despite the resentment he faced, Grant completed an arts degree with a major in sociology and politics. His brothers soon enough also had various letters after their names, but Brian had to withdraw from his studies because he couldn't find the time amongst all his other commitments.

Brian's stance on education benefited all four boys, whose qualifications eventually presented them with more opportunities. At various times they moved regularly between operational and administrative roles. Nowadays, their résumés read as though they are police service directories: communications; social justice; traffic enforcement; disaster management; counter terrorism; public sector management; legal policies; juvenile aid; curriculum development; ethical standards; crime and misconduct...

All the Pitman men have been heavily involved in educational roles, including the development and implementation of training programs. They have also taught at the Academy, presumably with arms anywhere but across their chests. For men who have built both their careers and lives on the desire to help others, it's not surprising that they gain great satisfaction from imparting their experiences.

Much of their knowledge has come from the operational field. Each brother has very poignant and powerful memories that reflect as much about them as some of the unfortunate people they've come in contact with. Grant recalls the day a drug dealer thanked him for being so polite during the charging process.

There was also the occasion that he received a lottery ticket from the parents of a woman who'd been seriously injured in a car accident. It was their way of thanking him for his compassion. However, he didn't accept the gift because gambling is against the Mormon faith.

Garth recounts the story of a 'young fellow heavily into drugs who was stealing motorbikes and always in some sort of trouble'. He was being questioned, yet again, when he looked at Garth and said:

'What can I do to change things?'

'Change your friends, change the way you think. Find out what you really want to do,' replied Garth.

The man did try to rebuild his life, and eventually sought a new start as a station hand in the country, although word got back to Garth that he reached the bush on stolen wheels.

'Some people at least make an effort,' says Garth. 'Every police officer knows families that have experienced the death of someone close, or endured a tragedy, such as a child on drugs. The lights of hope go out in people's faces. I personally find the loss of hope the hardest thing about policing.'

Stacey also has thought-provoking accounts. He vividly remembers the day he saw a small child's body on an autopsy table. He'd been in the mortuary many times before, but the fact that he had a child of similar age created an emotional link. The fate of that child stayed with him.

'The other one that really hit me is a house we were called to. No one was home, but the door was open, and the state of the inside was appalling. The sink was putrid, there was no clean floor space, and there were kids' toys scattered through the mess. It was a house that you could drive past every day and not give a second thought to. It really made me think. There are stories behind every wall in every house, and not all of them are happy. It made me realise how lucky I was, and still am.'

The beliefs and attitudes of Grant, Garth, Kendall and Stacey are wrapped up in a story Kendall tells.

'During the course of an investigation I asked a suspect why he was on drugs, and as with so many users he told me he really wanted to clean himself up but he was dependent and didn't know how he could do it. I told him he could come and talk to us [the police] at any time and we would put him in contact with agencies that could help him. About two months later, he showed up at my home. I don't know how he found me. When I saw him I immediately went to my wife and said: "Lock all the doors, lock all the windows. I've got a guy out there and I don't know what he's likely to do." Well, he was looking for help. He'd made the decision that he wasn't going back to live his life the way it was. From that meeting I was able to introduce him to an organisation that would look after him: his drug issues, his accommodation, everything. About six months later I received a phone call from him. He said "Thank you, you've changed my life." You know every time someone asks the police for help, they believe in us, they believe we can do something for them. That is a wonderful feeling.'

It is obvious the Pitmans are 'people's people' who genuinely care for those around them. This starts with their families. Grant has six children and two grandchildren; Garth has four kids; Kendall two; and Stacey three. Only one of this next generation is a police officer, but there are two others in the extended family who are aptly named brothers-in-law because they have joined both the Pitmans and the Queensland Police Service.

To an outsider, it is easy to suggest that devotion to a religion doesn't necessarily blend well with police work. But for the Pitman family, the two are intrinsically linked. Grant, Garth and Kendall have all been ordained bishops of their Church and, together with Stacey, they are all highly commissioned police officers. Whether they dress in their Sunday best or have epaulettes on their shoulders, they wear uniforms that are woven with the rich thread of humanity. It's this thread that ties their lives, beliefs and occupations together.

Much guidance came from their father, who retired from the Police Service on his 55th birthday, although on the morning that he said his goodbyes he contacted Kendall and pleaded: 'Can you go in and tell them to tear up my papers. I don't want to go just yet.' He had reached the rank of superintendent. Sadly, Brian now suffers from Parkinson's disease and is restricted to a wheelchair, but his police presence lives on through his sons.

Grant, Garth, Kendall and Stacey Pitman have shaken thousands of hands, and cuffed many others. For four brothers well versed in delivering messages and taking statements, one sentence is all that's needed to underline their characters. It comes from Kendall: 'We will never give up on people.'

JULIE ELLIOTT

'Friendships'

'Life can be bitter and sweet. It's the people around that you love that make it so beautiful.'

Senior Sergeant Julie Elliott, Queensland Police Service

SUNDAY 7 DECEMBER 2003

Thirteen-year-old Daniel Morcombe stood waiting at the bus stop on the Nambour Connection Road under the Kiel Mountain Road Overpass. It was only a kilometre from his home, and a little further than that to the Big Pineapple, one of the Sunshine Coast's most famous landmarks. At that moment Daniel had seen enough fruit for the day. He'd spent the morning picking passionfruits with his twin, Bradley, and older brother, Dean. Now it was time to go and do some Christmas shopping and have a haircut. It was about 1.30 pm; not long for him to wait for the 1.42 service heading to the Sunshine Plaza. But Daniel didn't know that the bus had broken down, and two replacements were on their way. One was to pick up the people on the broken-down vehicle, and the other would collect new passengers. It was after 2 pm when he saw the first bus. He hailed it, but it didn't stop. Minutes later, the second bus drove by. The driver had no reason to pull over. Daniel was gone.

MONDAY 8 DECEMBER 2003

Senior Sergeant Julie Elliott watched the blue eyes and cheerful smile stretch across her computer screen from the opened email attachment. Her instinct immediately told her that something wasn't right. This wasn't the picture of a street-wise boy who would run away from home. This boy's face was very sweet and innocent. This was no ordinary Missing Person. Julie had two overwhelming thoughts: 'The poor boy, and the poor family.'

***

When Julie Elliott was a recruit at the Queensland Police Academy, she was affectionately known by her classmates as 'Mum'. It was an obvious nickname for a 34-year-old woman who had sons aged five and seven. But if her young colleagues knew the journey she had taken to reach this point, perhaps they would have tagged her 'Herb' in reference to the great Australian middle-distance runner and Olympic gold-medallist, Herb Elliott. It wasn't so much the surname that linked these two, but the miles they had covered. By the time she was enjoying the Academy banter, Julie had rounded many a bend to achieve her long-term aim of becoming a police officer. She says: 'Police work always attracted me. I watched a lot of police shows when I was younger, and although there was a lot that wasn't like real life, I got an idea of what they did.

And I liked what they did. I always thought I could do it, but a few other things popped up along the way.'

Julie was born of blue-collar blood, the younger of two children. Her father poured molten metal at the BHP steelworks in Newcastle, while her mother poured beers in pubs. Although they had 'bugger all' as children, Julie and her brother didn't know they were missing much until there was reason to look. And that happened the day their mother walked out on the family. Julie was still a young girl.

Five years later, having departed from her childhood long ago, Julie left school and became a trainee bank-teller. It didn't take her long to realise a caged environment wasn't for her, so she moved on. She washed cars, packed fruit and, in her favourite role, she worked on her tan as a beach bum.

By her late teens, Julie took a surprising turn when she began nursing at Stockton Hospital, an institution for the mentally ill and developmentally delayed. Death, distress, and an overwhelming sense of depression were never more than a few corners away. Julie coped by partying hard. But after a few years, she needed more of an escape than drunken nights on the dance floor. The job had begun to daunt her; it was time to leave the thankless twelve-hour shifts to someone else.

She was in her early twenties when she headed to the Gold Coast with a group of friends and no fixed plans. They only booked a flat for a week, but stayed for a few years. Julie found work in a guesthouse, and as a restaurant waitress. Life was simple. Life was good.

The wanderings of her early adulthood eventually led her to Brisbane where she met Dennis, a man who also had gypsy feet. They fell in love and, in the blinding mist that romance brings, they decided to travel around Australia. They bought a Land Rover, a boat and a trailer, and headed off into the unknown. After months of camping, shuddering over roads, and slapping the boat into both still and rolling waters, they reached Albany, a coastal town 400 kilometres south-east of Perth. They were so charmed by the area and so tired of changing gears that they bought a cottage for $6500 after noticing its picture in a real estate agent's shop window. They opened a cane furniture business and also sold fence posts, which they made from felled trees. The latter process was simple. Dennis and a partner cut the trees down, and then blew them apart with dynamite before Julie stepped in to strip the bark. Two years later they returned to Brisbane, got married and started the 'baby process'.

They had two sons, Paul and Mark, born two years apart. Julie thought that she was 'the cleverest person in the world'. Apparently there were a few other women who were mothers too, but this didn't stop Julie basking in the brilliance of her achievement.

In search of a better family lifestyle, Mum and Dad and their two boys moved to Caboolture, about 40 kilometres north of Brisbane and at the fringe of the rapidly developing Sunshine Coast. While Dennis worked in the telecommunications industry, Julie combined motherhood with a number of jobs, including one as a teacher aide and administrative assistant at Peachester State School in the hinterland. She was there until she noticed an advertisement in the paper. The then Queensland Police Force was looking for new recruits. Julie spoke about this opportunity with Dennis, whose support she needed if she was even to try to get in. Although their relationship was having some troubles, she applied. Within a few weeks she walked nervously into a room at Police Headquarters in Brisbane. Five men sat behind a long desk. Four were commissioned officers, the other a psychologist. The room's only vacant piece of furniture was a chair that was pushed against a wall. It was several metres away from the desk, leaving a sea of intimidating space. Julie immediately picked it up and moved closer to the desk.

'No, we want you to sit back there please,' said an inspector.

'No, I'm not comfortable there, thank you. I'd like to be closer,' replied Julie.

The inspector laughed.

'You're in! he joked.

But earning acceptance wasn't as simple as lifting a chair. In the eyes of the panel Julie also carried her own hurdles, chiefly her age and her children. At the end of the intense session, she expected to walk out the door and back into her normal life, but that changed with just one word: 'Congratulations.'

Only a month later she began one of the biggest challenges of her life. This mother of two was 'Mum' to dozens of others at the Academy. She was the oldest recruit in a squad of 25. But returning to study for the six-month course was extremely difficult for Julie. Because the Brisbane-based Academy was a live-in facility, Julie was forced to live away from her family for the first time. She only went home at the weekends. Although she was there to cheer her sons at nippers and rugby, she missed the everyday interaction of a mother and her children. During the week, her contact with home was limited to phonecalls, which often ended with a sniffle or tears from her, not her 'babies'. It was bizarre how life could throw twisters to test the character. So many years earlier, Julie discovered what it was to be a child missing a mother; in her adulthood she discovered what it was to be a mother missing a child.

Although she could never replace the time away from her children, 'Mum' found her Academy surrogates were delightful company. Among her closest friends was Lance, or 'Buddha', a short man in his mid-twenties who was carrying a little extra weight. He was living his dream of becoming a police officer. It was all he ever wanted to do. On Lance's very first day at the Academy he endeared himself to his classmates when he told them how worried he'd been that his lack of height would hamper his chances of being accepted. However, he'd read somewhere that people could stretch if they lay down for 24 hours. He did this prior to his interview, and obviously looked suitably tall enough to impress the panel.

While Lance might not have been able to see the tops of some people's heads, he could certainly see inside them. A few months into Academy life he knew Julie was battling without her children, so he passed the hat around the squad and raised enough money to hire a bus for Paul and Mark to come to visit on a school excursion.

Despite, at times, struggling with the Academy's rigorous physical challenges, Lance graduated with distinction with the rest of his classmates, including Julie, whose commitment was rewarded when she was voted as 'Best All-Rounder', a prize for the recruit who had most impressed at both the physical and mental sides of training. After being told of his posting in the final week of training, Lance knocked on Julie's door with a six-pack of beer in his hand.

'We've made it!' he said.

Little more than a year later Lance died from leukemia. He was one of the first truly 'special people' that Julie met in the service.

Although Julie had covered a vast distance by the time she became a police officer, many more miles and bends lay ahead. Constable Elliott was first posted to Maroochydore on the

Sunshine Coast, but within a month she was working at Caboolture. Home at last.

When she was on day shift, Julie's sons would often go to the station after school and wait in the kitchen, occasionally poking their heads out with wide eyes when a 'baddie' was taken past. They also loved being picked up at school by Mum when she was in uniform; this earned serious playground prestige.

During her years working at Caboolture, Julie had the typical experiences of a general duties officer: break and enters to common assaults; drink-driving to public mischief; domestic disputes to murders. She became involved in community programs, and was an Adopt-A-Cop at a local school and a War Veterans' home, where tea flowed as readily as opinions on how to solve crime.

She knew it was time to move on when her job began standing toe-to-toe with her personal life. Occasionally when shopping on a day off she would hear venomous whispers behind her: 'You bitch! You pinched me last night.' She accepted those moments as part of the job, but her thoughts changed the night she and three other officers were called to a hotel, where a group of men were smashing furniture. They managed to arrest the main offenders and take them back to the station. This triggered an explosion of rage. Within minutes, others arrived outside the station with baseball bats and knives. The mob smashed windows, slashed tyres, and screamed abuse. Julie and her colleagues were forced to lock themselves in, and call for back-up. While other units eventually arrived to combat the street fury, inside the station a stare and a statement cut through Julie. She was in the watch-house helping her sergeant process the charges when one of the offenders spat out the words: 'I know where you live. You better keep your doors and windows locked, you bitch!'

A string of profanities followed. Although she tried not to show any emotion, Julie was unnerved. Soon afterwards she transferred to Brisbane, where she worked in the recruitment division. She was clicking over the miles again, visiting universities and TAFE colleges across the state, and believing passionately in what she said to the students: 'I love being a police officer. It's a great job, and you really can make a difference.'

For seven years she periodically shifted between recruitment, operational and training roles, including a teaching stint at the Academy. But she too was still being taught, as she was constantly exposed to myriad lessons of life. Violence, laughter, abuse, survival, tragedy, relief, tears, celebrations, conflict, happiness, death . . . all were themes that could sometimes swirl together with a degree of explanation, but not a depth of understanding. As a police officer it was important to keep a professional distance from some people and emotions, but how was an officer to react when the issue was personal?

When Dennis first began behaving strangely, Julie assumed her husband was simply having an off day or two. But those days soon grew longer. A burst of tears could rub shoulders with laughter; clear logic could surrender to paranoia; joy could be shoved aside by hopelessness . . . Bipolar (manic) depression is a horrible illness.

There were rough times and Julie and Dennis parted. Paul and Mark stayed with their mother. The boys hoped that someday they would all get back together again as a family, but that hope was locked away forever the day Julie arrived at Dennis's home to find she couldn't open the garage doors.

'I knew I'd find him dead,' she says softly. 'He felt he had to escape his demons, and everyone would be better off. I've never been angry with him. It's a strange feeling having joined the police, gone to suicides, and then one day finding you're in the middle of one as a person not an officer. I'd been to ones where I'd heard people say: "Oh the weak bastard, the coward," but I'd always say, "You're not in their shoes. And I hope you never are." The hardest thing I've ever had to do is tell the boys what had happened. That will have an impact on their lives forever. I think the biggest thing I learnt out of it all was that there is nothing ever too terrible that you can't talk to someone about it. You have to find strength from others as well as yourself.'

Julie and her sons fought through their grief and gradually built lives that would accept but never forget the past. As Paul and Mark headed towards adulthood Julie became less a mother and more a mate. She took each of her sons to the pub and shouted him a drink on his eighteenth birthday. The same was done for their 21sts. They were simple but telling gestures from a woman who truly understood 'there is no love quite like a parent's love for their child'. At that stage she couldn't have known that those few words would have such a powerful meaning when her career took another change of direction.

In 2002 Sergeant Julie Elliott was one of three officers placed in trial roles in the Queensland Police Service Media Unit in Brisbane. Until then only civilians, chiefly former journalists and public relations consultants, had managed this area. Julie enjoyed the challenge of learning new skills. She answered inquiries, wrote releases, arranged interviews and all-in media conferences. She liaised with interviewees, both civilians and police; she trained officers in and worked on strategies that were used when there was a need to arouse public attention or gain information about particular issues or incidents. After six months the trials were over, the officers stayed, and Julie settled into a job which may have taken her away from the front line, but exposed her to more cases than any officer would ever see in a whole career on the beat. All the miles, all the bends, all the ups and all the downs had led Julie to this point. And all had combined to teach her that the biggest certainty about life was its uncertainty.

She had been in the media relations role for more than two years when that uncertainty became the centre of her life. She was nearly 50 years old, and had spent her last fifteen as a police officer. Despite all her experiences, nothing could wholly prepare her for the days, months, and now years that followed 7 December 2003. After she saw the image of 'beautiful sweet Daniel' appear on her computer screen for the first time, she soon had reason to wish, as a police officer, a mother, a human being, that uncertainty didn't exist.

***

When Daniel Morcombe was born, he and his twin brother could fit together in an adult's hands. Neither of the tiny babies weighed four pounds, which prompted their mother, Denise, to jokingly name them 'Number 15 and Number 16', in reference to supermarket chickens that were of the same weight. With their brother, Dean, two years their senior, the boys grew up in a Melbourne family that was full of love. They eventually moved to the Sunshine Coast, where Denise and her husband Bruce established a thriving business in landscape gardening. Water and earth were synonymous with the Morcombes' way of life. The surf and sand were close, the soil and sods closer still. In this sometimes peaceful, sometimes hectic environment, Daniel matured into a teenager whose quiet side was reflected by the collection of goldfish that he and Bradley nurtured. In contrast to this, Daniel could assault the ears, never more than when opening the throttle and throwing himself into the air on his motocross bike. Whatever he did, he was simply a typical teenager finding his way towards young adulthood.

The day Daniel disappeared was just twelve days before his and Bradley's fourteenth birthday. With school holidays and Christmas also approaching, it was a great time of year.

But then he went missing.

He was last seen at about 2.10 pm on the Nambour Connection Road under the Kiel Mountain Road Overpass. He was wearing light-coloured shoes, white socks, dark knee-length shorts, and a red T-shirt with 'Billabong' printed across the chest.

Two days later, Julie was meant to travel to the Sunshine Coast to give a series of training presentations to officers. But with investigations intensifying into Daniel's disappearance she instead found herself organising a media conference that was to involve the Morcombe family.

After all the standard preparations had been made, she knocked on the door of a room at the Maroochydoore Police Station where the family was waiting prior to the media conference. She entered into a 'silent, stunned state of shock' knowing that all the people who were suddenly staring at her were desperate to hear words that she couldn't say: *'Daniel has been found.'* She introduced herself to Bradley, Bruce and Denise, and Denise's parents, Kevin and Monique, who'd flown from Melbourne. Dean was with friends. There were two other police officers in the room: Senior Constable Samantha Knight, a friend of Julie's, was in a special liaison role; and her boss Senior Sergeant Greg Daniels, Officer in Charge of the Juvenile Aid Bureau, was also present. Julie explained what would happen at the media conference. Denise said nothing. She couldn't. But Bruce spoke firmly: 'I don't cry. I'm not performing.'

'No one is asking you to do something that's not natural to you but we've got to take advantage of the media that's here,' replied Julie, who had typed out some bullet points on a piece of paper that she gave to the family. 'You can read them if you like, but the message has to be clear. You want information. You are worried.'

It was a horrible way for people to first meet, but the Morcombes acknowledged that Julie needed to be frank. In these circumstances it was critical to obtain public information as quickly as possible.

At the media conference Bruce did as Julie and her colleagues had hoped. He spoke clearly, controlling his emotions in a room packed with journalists. Denise, however, barely said a word. She sat next to her husband and stared ahead, and at the floor. In the absence of her son, she clutched on to hope.

There was strong media interest. In the harsh world of headlines this case had ingredients that would make a big lead story: a loving and respectable family; twins; a perceived abduction in broad daylight; and the mystery of the disappearance.

Momentum gathered quickly as reports flooded in to the police. Such was the volume of information and inquiries that Julie was quickly taken away from her regular duties and appointed wholly to the Morcombe case. Every single media inquiry had to be dealt with by her. This included requests for interviews with the Morcombes, which meant Julie was in contact with the family every day. She relayed the requests carefully, talking through the specifics of each interview: where it could be held; what photos or footage would be needed; what was the angle of the story; how long it would take. In essence, Julie began to manage the Morcombes' public lives. A quiet family had been thrust into the spotlight, a position it knew nothing about, but had to accept, no matter how great the pain.

The Morcombes' private lives also changed as they were forced to confront issues that they thought only unfortunate others had to contend with. But this time the Morcombes were the unfortunates. Because of the nature of the case the police needed to dig into potentially painful subjects. Were they having money problems? Was someone having an affair? What type of boy was Daniel? Could we take DNA swabs please?

Julie too had to 'cross the boundaries' in a role that demanded both compassion and a hard edge. A week after Daniel went missing the police staged a re-enactment using a mannequin at the Kiel Mountain Road Overpass in the hope that it would trigger public reaction and the memories of passers-by. For full impact in front of the media, the police knew it was time for Denise to speak, but she told Julie she couldn't.

'Yes, you can. And you will,' replied Julie. 'It's Bruce's turn to shut up. The media wants to hear from Mum now. You have to do this.'

Julie and Denise quietly rehearsed what needed to be said, and then Denise was on her own in front of the cameras.

'It's getting desperate. We need him back. We want Daniel back,' she said, fighting back tears.

As Julie watched on, she was torn between two roles. On one side, she was trying to imagine as a mother 'how horrible, how awful' the situation was, but on the other side she was a police media officer who saw the scrum of journalists and cameras as 'great coverage'. And when Denise began to cry,

Julie thought to herself: 'Thank you'. Tears made powerful pictures.

Despite all the media attention, exhaustive police work, and overwhelming community support, the days passed without any clear breakthroughs. The best lead concerned a square-style 1980s sedan with faded blue paintwork that was seen near the overpass at about the time that Daniel went missing. There were also witness reports of a man standing behind Daniel at the bus stop. He was described as being 'between 25 and 35 years old, about 175 centimetres tall, lean to muscular build, thin gaunt face, dark brown wavy hair, goatee beard and with a weathered complexion'.

More days passed. Frustrations and desperation grew.

If fate could offer a kind hand, 19 December was the day to deliver news that Daniel was alive and well. It was his fourteenth birthday. Somehow, amid the torments of uncertainty, the Morcombes still had to acknowledge Bradley's birthday. It was going to be a day so taut with emotions that stoicism could slip into grief in the mere flicker of a thought. It was a private time for the family, but in the eyes of the police it was a private time that needed to be shared with the public. So Julie, who was by now very familiar with sensitive intrusion, asked the Morcombes the question that led to her feeling like 'an evil part of the day'.

'I know this is a very big request, but could we please have media access for Daniel's and Bradley's birthday?'

Denise and Bruce consented. Despite the prospect of revealing their rawest feelings they understood the importance of maintaining a high level of coverage for the case. Julie arranged a media pool, which included a television cameraman, a photographer, and a radio journalist who represented the industry as a whole. All footage, photos, interviews and information were to be made available to every media organisation. This wasn't the time for exclusives.

The day arrived in what Julie now considers was a 'turning point' in her relationship with the Morcombes. Together with the media pool, Senior Constable Knight and Senior Sergeant Daniels, she wasn't just allowed, but was welcomed into the Morcombes' closest unit of family, friends, and two Catholic fathers. They sat in the shade of a tree in the Morcombes' garden. Three photos stood and smiled on a table next to three unlit candles. Julie had shaken hands with two of the brothers in the prints, but she best knew the third, although she feared she'd never meet him. In the glow of the late morning a breeze whispered while the people gathered prayed. There were hugs and tears. Speeches and tears. Silence and tears. Then Bradley lit the candles. The flames of two burned brightly, but the third flickered weakly before going out. A short while later, Denise's mother Monique told Julie: 'I know my Danny is dead.'

But Denise and Bruce weren't ready to accept the worst. In a bid to create further public awareness Bruce found an old white door with a key in it. He wrote on it in bold letters:

WHERE IS
DANIEL?
WHO TOOK HIM?
HELP FIND THE
KEY TO UNLOCK
THIS 'HORROR' OF
A CRIME
WE WILL NEVER GIVE UP
The Morcombe Family

He also tied a red ribbon to the door's handle to signify the colour of the T-shirt that Daniel was last seen in. Then, he placed the ingenious billboard on posts at the bus stop where Daniel had vanished.

Despite a constant flow of information to the police, the door remained locked, and Christmas passed by the Morcombes' house without knocking. By this stage, Julie had walked across the threshold many times, and her relationship with the Morcombes, particularly Denise, was beginning to change. They had graduated from polite handshakes to tentative pecks on the cheek, and then to hugs. With the physical barriers broken, the emotional ones soon followed, as conversations spread from strict discussions about media commitments to wandering reminiscences about Daniel. A high level of respect and trust had been there from the first day Julie and Denise met, but now a professional relationship was growing into a friendship. This strengthened early in the New Year when a production team from ABC Television spent several weeks producing a program about Daniel's disappearance for the widely respected 'Australian Story'. The sheer logistics of the production meant Julie was rarely away from the Morcombes during this time. She ate with them, drank with them, and even stayed the occasional night in their home instead of making the drive back to Brisbane. Julie had become more than a police officer to the Morcombes, and much more than a friend to Denise.

'I really love Denise and this family,' says Julie, beginning to cry. 'They were, and still are, the most amazing people. Amazing people! They'd never spoken to a journalist before in their lives, and they certainly hadn't had to deal so closely with the police. And then suddenly the police and media took over their lives. They accepted it, and even in their grief, although it was something they hated, they knew there was a purpose to everything they had to do. Denise and I had become very close. It was impossible not to because she was just so easy to love. On any other day it could have been me, or another mother put in her position. Except for those people who are forced to go through it no one could ever understand what it must be like to lose a child in this way. It was so awful. I just felt so sorry for them.'

The days and weeks dragged into months, and still there was no word on Daniel's whereabouts. Since their lives had been torn apart, the Morcombes had shown incredible bravery and strength, which had earned them admiration and sympathy throughout Australia. Thousands of posters with Daniel's photo and details of his final movements had been distributed to schools, police stations, shopping centres and businesses across the nation. Two 'Red Ribbon Days' had also been staged. The first was on the Sunshine Coast, the second spread across Queensland after it was endorsed in Parliament by the Premier Peter Beattie. School lockers, homes, fences, letterboxes, power-poles, shopfronts, lapels, and even car aerials wore the simple looped strip of red that had not only become synonymous with the case, but was the symbol of a state-wide campaign to protect children against crime.

In early May 2004, Julie and the detective inspector in charge of the case, John Maloney, accompanied Bruce and Denise on a trip to hold interviews and media conferences in northern New South Wales. This followed speculation that the blue car that police were seeking may have had New South Wales numberplates. It was a journey of emotional torture in which new miles were travelled, and old steps re-traced. After one exhausting day on the road, Julie and Denise sat together in a Coffs Harbour hotel room. Bruce and Inspector Maloney had already gone to bed. It had been nearly five months since Daniel had gone missing. Five long months. Finally, the time had come for Denise to confess: 'I know he is dead,' she said quietly.

She wept. Julie did too. They held on to each other and sobbed.

In the following days Julie spoke of her own situation with Dennis and his suicide. The knowledge she had gained from that was now being passed on. The two women, drawn together by tragedy, had both endured personal sufferings that bonded them in a sad but powerful way. They spoke frankly with each other, and even 'could get angry with each other without feeling we didn't deserve to be roused on'. They drew strength from one another.

As more time passed, the Morcombes tried to accept life without Daniel. Julie returned to general media duties but she remained in close contact with Denise. Occasionally they stared through an empty bottle of red wine together. In August 2004, Julie walked the Kokoda Track in Papua New Guinea to celebrate her 50th birthday. Although she relished the experience her thoughts were never far away from the Morcombes. If Daniel was found, Julie 'wanted to be there for Denise'.

On 7 December 2004 a memorial service was held for Daniel. Afterwards, Denise and Julie made a pact. A week later, they lay screwing up their faces in a tattoo parlour. Denise's right ankle was coloured with a red ribbon under which 'Daniel' was written; Julie had an entwining pattern etched into the small of her back. Life has since twisted on for both women who continue to have a relationship that reflects every light and shade of emotion.

At the time of writing, Daniel still hasn't been found.

'I would love to have known this child,' says Julie. 'I feel I do. I feel I know everything about him. I look at Daniel, and I know his pain is gone. I believe he is dead. I don't like thinking about what he might have endured. Denise has asked me that too many times. Things can be taken away so quickly. An accident is an accident, and your mind allows you to balance that in time, but to think that some evil bastard in all likelihood decided to pluck this decent young man off the street and put all these people [the Morcombes] in a living hell is totally abhorrent. That is why we are in this job. We still have to do everything possible to make sure we find Daniel so his parents can bury him and say goodbye.'

Julie will never stop hoping that one day the uncertainty will be over. No matter what happens in the future she will always offer Denise and the Morcombe family tremendous love and support. A 'different love' is reserved for her own family. In her 51 years, her most heart-warming and heart-breaking experiences have taught her that nothing is more important than the happiness of her sons. She shares drinks with them, trades insults and jokes, and has even been known to draw back on the odd cigar. And when she does she thinks: 'This is wonderful. I am so lucky.' She is also 'gaining great strides' in a renewed relationship with her mother and understands more clearly now why some choices were made many years ago. Sadly, her father has died. She acknowledges: 'My life has sometimes been a bloody hard road. I've aged a lot but learnt a lot about myself too. I've got a lot more capacity to give and take than I thought I ever had. I've been stretched the odd time and just when I thought I had nothing left, I'd find the strength from somewhere to go on. Everything has taught me that I must be happy, and it's important not to waste opportunities. You must treasure life.'

And for those who know Senior Sergeant Julie Elliott it's also important to treasure 'Mum'. Her journey is one that can make us all stronger.

IF YOU KNOW ANYTHING THAT COULD HELP THE INVESTIGATION INTO DANIEL'S DISAPPEARANCE CONTACT CRIME STOPPERS ON **1800 333 000** or EMAIL crimestoppers@police.qld.gov.au AT ANY TIME.

INFORMATION CAN BE PROVIDED ANONYMOUSLY. CASH REWARDS MAY BE AVAILABLE FOR CERTAIN INFORMATION.

NEALE SMITH

'To Reach Great Heights'

'Life is finite. There is no guarantee you are going to be here tomorrow. You have to value where you are at in life. You need to appreciate your family and friends and your place in life with them.'

Paramedic Neale Smith, Tasmanian Ambulance Service

Rock-climbing can be addictive. To those who are hooked, there are challenges which are like the tiny foot-holds and finger-holds that always seem just a little out of reach. You cling to the frown lines in a haggard cliff face, and work your way up or down, inch by inch, thought by thought. You may need to push yourself to the very edge of your mental and physical limits. It is a test of your character, skill, determination. The strain on your fingers, toes and limbs; the contortions of your body; the twists in your mind all have to be overcome. And when they are, there may be an overwhelming rush of excitement or a simple breath of relief that trails a few simple words: 'I made it.'

Neale Smith knows that feeling very well. Ever since he was a boy building forts and playing wars in the bush with his brothers and friends, he has had a connection with the outdoors. He was raised in Roland, a tiny village on Tasmania's north-west coast, an area of haunting beauty. Neale needed only to walk outside his family home to feel as though he could reach out and touch the heart of the district's landscape, the imposing Mount Roland. When he was a teenager, he trod the trails at the mountain's feet, but as his quest for adventures grew, he veered away from common paths to rock faces that threatened most passers-by, but dared others. By the time he'd left school, he and his climbing mates spoke about such faces in the same way that other young men referred to the football grounds they'd played on.

'Climbing pretty well became the focus of my life. We went out and did new climbs, discovering new cliffs and first ascents anywhere in Tasmania. Some of the climbs were very hard, so we were pushing ourselves. It's different from climbing now. What we were doing was traditional climbing. We'd start at the bottom, work our way up, and place our own protection as we went. In some of the harder ones we'd fall off and have to work our way up again. There's the mental side you have to really overcome. I don't like using the word fear; instead it's a definite challenge in working through mind sequences. It can be quite complicated working through all the puzzles and pieces. But it's also exciting. The sense of achievement gave us a real buzz. That drove us.'

During these years Neale was a winter mountain-craft instructor. He took groups of people out into the wilderness for a few days at a time, and taught them survival techniques, and practical skills. Some of his colleagues were ambulance officers who chatted so enthusiastically about their jobs that

Neale developed an interest in a profession that he'd previously known little about. When he was 21 he applied for a position in his then home town of Devonport. He was accepted, despite the fact that he didn't even have a first aid certificate. It was 1982.

In a bid to increase numbers in the Tasmanian Ambulance Service, Neale was fast-tracked through training and became a paramedic within three-and-a-half years. His bush and mountain-craft knowledge added to his value in a state where wilderness incidents were common. He spent twelve years in the area, performing any number of roles including rescue, relief work, volunteer training, and clinical instruction. In 1995 he moved to Hobart, and was soon accepted into the Wilderness Rescue Helicopter Squad which was a pool of local helicopter pilots, Police Search and Rescue officers, and paramedics. There was no formal training as such: it was simply a matter of 'if you were on shift when a case came in, you would go'.

As with climbing, the role of working in and out of a helicopter posed—and still does pose—constant challenges. Neale says, 'You go out of the chopper by yourself, down to what may be the unknown condition of a patient. Sometimes you may be winched down a couple of hundred feet. You possibly have a stretcher with you, some medical equipment and a spinal splint. Then you detach from the helicopter line, the chopper will wait for a quick assessment—like the patient's condition, how long will an assessment take or how long it will take to get the patient onto a stretcher—so during that time you're kind of left there by yourself. You're part of a team, but you also work alone. All the while you're reassuring your patient and managing them in what can be some pretty uncomfortable positions like a narrow ledge. Then you get them on the stretcher or take them up and through a small door that's 60, 70, 200 feet in the air. It's challenging. There have been times that I've had a real link with the patient. I appreciate what they've been doing out in the wilderness. I think "it could have been me". I empathise with them.'

And so it was on Friday 13 February 1998. Neale was nearing the end of a day shift, and already thinking about enjoying a quiet beer when his supervisor told him 'We might need you to go to an incident on a cliff.'

He was soon on a chopper heading 100 kilometres south-east of Hobart to a treacherous spindle of rock that, to this day, lures some of the world's best climbers. To conquer this structure is a treasured prize even for the elite. The aptly named Totem Pole is 70 metres tall and just 3 metres wide in most places. It stands alone 20 metres away from the sheer, spectacular cliffs of Cape Hauy that overlook the Great Southern Ocean. Even on a pleasant day, the waves here can be frightening as they slam into the rocks and spit high above the waterline; on an ugly day, when southerly gales howl in from the Antarctic, it needs no imagination to realise what could happen to a person trapped by nature's fury.

Thankfully, it was a calm evening when Neale and police officer Sergeant Paul Steane considered what lay ahead. They'd been told a man was seriously injured somewhere on the Totem Pole. His girlfriend had run out along a track and informed a park ranger of the situation. The Police Search and Rescue team had been informed, but it would still be some hours before they'd arrive at the scene, because they would have to walk in along several kilometres of bush trails with their equipment. It would then take them further time to set up. Neale felt it was going to be 'a long-drawn-out affair'.

When the chopper arrived, Neale looked down at the sea stack that he had never climbed. He had thought of doing it, but he needed time to practise; time that in the previous few years had become increasingly rare to find for a husband and father with a three-year-old son.

Despite its daunting reputation, the Totem Pole was dwarfed by the cliffs of Cape Hauy and another pinnacle, the Candlestick. Two men were on a ledge about 25 metres above the water. The injured one was lying down, and the other was sitting next to him. Neale, Paul and their pilot initially thought they could try hovering above and winching down, but it was too dangerous a squeeze next to the cliffs, and the rotor-wash would create further problems. They decided it wasn't an option.

They flew to a point about a kilometre away from the cliffs. They presumed they would be able to do nothing more than walk to the point of Cape Hauy, assess the situation by peering over the cliff's edge and then wait for the Police Search and Rescue team. The clock was ticking.

They reached the Cape and, after walking cautiously to the edge and looking over, they noticed a Tyrolean rope traverse (Flying Fox) stretched across to the Totem Pole's summit from a point on the cliff about 30 metres below them. They scrambled down to it. Neale looked to where the injured man lay with his helper on a ledge.

'Can you hear us?'

It was impossible to know if any shouts were coming back. Distance and the sound of crashing waves made voices disappear. It was 7.45 pm; in little more than an hour it would be dark. Neale began to wonder: stay put or have a go? He inspected the anchors of the flying fox. Yes, they seemed solid enough. Should he? Shouldn't he? He again looked at the injured man below, and after a quiet moment of contemplation he decided, 'Bugger it, I'll go across and have a look.'

He had some basic climbing equipment: a harness, a couple of karabiners, and a figure of eight descender. He clipped himself into the ropes and edged his way out. He was 90 metres above the water, pulling himself across hand over hand with his back facing downwards and ignoring oblivion. Any experienced climber knows the immense strain that is placed on the anchors of a flying fox. If the pressure is too great, there's the chance that the anchors will be ripped out, often from only one end, sending the climber hurtling through the air still attached to a rope and swinging towards disaster. But Neale wasn't afraid as the waves licked their lips below. His thoughts were already focused on what he would find at the end of his perilous trip. He reached the Totem Pole and stood for a moment on the summit. The area of just a few square metres made Neale feel as though he was swaying on top of a pinhead.

Moments later, he was abseiling towards the narrow ledge that held two other men, and one fading life.

He was greeted by Tom Jamieson, one of two Australians who'd met the injured climber's girlfriend, Celia Bull, when she was running for help. While one hurried off with Celia, Tom went to the Cape to see if he could do anything. The accident had happened several hours before, when the climber, after making his way across the flying fox and abseiling to the base of the Totem Pole, was preparing to start his ascent. He was hit on his head (he wasn't wearing a helmet) by a rock about the size of a small television that had been dislodged by a rope above him. By the time Celia reached him, he was bleeding heavily and fighting to remain conscious. Celia set up a pulley system and slowly, painfully slowly, began hauling her partner up, but after about three hours and 25 metres, she couldn't go any further. She then climbed to the top and rushed away to get help.

When Neale arrived his patient had a 10 × 5 centimetre hole in his head, and was lying in a pool of blood that covered much of the ledge.

'My first thoughts were that he [the patient] was unconscious but there were times when I thought he was able to hear me because I could get occasional movements like a squeeze of the hand. He had no movement on the right side of his body. The situation was grave. I looked at him and thought: "You've lost a lot of blood, and you've been down here for hours with a very bad head injury. If you stay here for any longer you are going to die."'

Neale tried shouting to Paul, but neither partner could hear the other. However, voices from below carried up from a State Emergency Services boat that was drifting close to the Totem Pole. It was a 6-metre aluminium runabout with a handful of SES personnel and police officers on board. In the relatively calm 1-metre swell it was possible that the vessel could make it right to the base of the sea stack. Although it seemed a rescue could be attempted, Neale still had to contemplate his position. He'd already placed a cervical collar on his patient, but had nothing else to secure the spine. Any movement without the proper equipment could perhaps cause serious damage, but when this possibility was considered against the growing likelihood of death, Neale had to take a chance.

Within minutes he'd set up a ropes and harness system that he hoped would allow him to abseil to the bottom with his patient secure across his lap. He tested his weight against the harness and after deciding it was strong enough, he dragged his passenger to the edge, waved to the men in the boat and stepped into the void. With his feet pressing firmly against the rock, and Tom holding a safety line from the ledge, he pushed nearer to safety, fighting to keep control. The SES boat moved into position. The waves leapt, the towers of rock seemed to crowd in as the last of the sun disappeared. Five metres, four metres, three, two, nearly there . . .

'Shit!'

The boat was pushed back by the swell. It came in again, out again. In again, out again. After all the dangers that had passed, why had nature chosen now to play a game? On the fourth attempt Neale lowered himself a little further and managed to hook his feet over the side of the boat, but as he did, the vessel began drifting away.

'Cut the rope! Cut the rope!'

A police officer reacted quickly, and Neale dropped into the vessel. In stark contrast to his patient he'd been on the Totem Pole for only twenty minutes.

They hurried off to a beach in Fortescue Bay a couple of kilometres away, leaving Tom on the ledge to be helped to safety later in the evening by the Police Search and Rescue team. As the boat ploughed through the water, slapping waves, Neale and his colleagues strained themselves to keep their weakening man from bouncing against the sides. They arrived at the beach to meet the rescue helicopter, and a volunteer ambulance crew which helped Neale place the patient on a spine board and load him into the chopper. They flew into the night and landed twenty minutes later at Hobart's Cambridge Airport where a paramedic crew was waiting to drive the final leg of this dramatic relay to the Royal Hobart Hospital. All the while Neale concentrated solely on keeping his critically injured patient alive. When responsibility was handed over to the hospital staff Neale had finished his job, yet he stayed around, waiting in the corridors, feeling as though he should be there. He finally left, and arrived home with his mind somewhere between a flying fox, a rock ledge and an emergency ward. As he drifted off to sleep, a stranger he'd come to know surrendered to anaesthetic. Englishman Paul Pritchard would be on the operating table for six hours, with his surgeon carefully picking out fragments of bone and rock from inside his head. Despite the seriousness of his injuries he would survive, but he'd face a long battle to walk, talk and think properly again. Nevertheless, he was incredibly lucky.

Neale had heard the Pritchard name mentioned at the hospital, but its relevance hadn't registered with him. It was only in the days afterwards that he realised he'd helped save one of the world's greatest and most daring rock-climbers. Over the following weeks he visited his famous patient. The two men soon became friends, and when Paul was released, it was fitting that he was wheeled out of hospital by the paramedic who'd begun the journey with him nearly five weeks earlier.

The two men still keep in contact to this day, which is easier than the difference in nationalities would suggest, because Paul was drawn back to Hobart by a woman he'd met during his rehabilitation. They are now married and raising a family.

In recent years Neale, too, has seen changes in his life. After making it through a divorce, he stood at the altar for a second time and has since become the father of two young girls, adding to his son from his previous marriage. Throughout his personal hardships and joys, his work with the Tasmanian Ambulance Service has been a constant. He is now a Clinical Support Officer whose primary role is in training and education. He attends a wide variety of cases, from wilderness incidents and helicopter retrievals to everyday occurrences such as road accidents. His knowledge matches his modesty . . . both are widely respected by his peers. He is a quietly spoken man whose loudest statements have been his actions. And nearly eight years after he took on the Totem Pole, his achievements still echo strongly in his memory.

'It was pretty much a blur at the time. I just went into automatic mode. I didn't think about anything much except a man who I thought was going to die. I related to Paul. He was a climber, I was a climber. I could feel for where he was at. You shouldn't have to justify why he was there. People should be allowed to do these things. When we were on the ledge I did think: "This could have been me or one of my mates." Clinically, I couldn't do much more than keep him as comfortable as I could and with the help of everyone else, get him to the hospital as quickly as I could. When we finally got to the boat the whole experience seemed surreal. It was almost like I was viewing happenings from outside myself. I guess if I look back, it was out of the ordinary, no doubt. It was one of those moments in your life when you genuinely do make a difference. I know that for Paul, it probably meant his life. It's a great feeling, you know. I consider myself fortunate to have helped someone in that way.'

Neale admits that 'it's possible one day I will go back to the Totem Pole and do it'. He will abseil to the bottom, then start the long climb up: foot after foot, hand after hand, thought after thought. If all goes well he will succeed, but in rock-climbing there can be no guarantees. Whatever may happen, there is one likelihood: the quest to reach the summit won't take him to the heights that he reached when he went down the rock on 13 February 1998.

GARY SQUIRES

'The Lifesaver'

'The beach was my front yard.'

Senior Constable Gary Squires, Victoria Police

Thirteen-year-old Gary Squires lay on his bed and contemplated the day ahead. He'd just been out to help his Nan and Pop do the grocery shopping, but there was little else planned. In a while, he'd probably go down to the beach with his board, or he might take his footy and have a kick with some mates, or he could go for a run by himself on the soft sand. Then again, he could just sink into his mattress and drift away on the lazy hours of the school holidays.

'Gary.'

Gary turned to see his Pop at the bedroom door.

'There might be some trouble. You should have a look.'

Gary leapt to his feet and hurried to the front of the house from where he could see the expanse of water that was his front yard. A minute later he was sprinting across the sand with a surfboard under his arm and three children in his sights. The nearest was about 100 metres away in deceptively calm water, the other two were double that distance from shore. All three were girls. Gary reached the water, his knees pumping. He was soon flat on his board, a banged-up piece of fibreglass that had swept across too many breaks. He began to paddle, focusing his eyes on the girl closest to him. She was already floating face down. After reaching her, he lifted her onto his board, and although barely able to stand up in the depth, he pushed back towards the shore, walking then running as his feet took a firmer grip. He saw his Uncle Peter and his Pop hurrying along the beach. Peter met Gary in the shallows and took the girl onto the sand where he began CPR.

Gary turned and headed back out, paddling towards the other two girls in trouble. He reached the first, who was struggling to tread water. He grabbed her.

'You're going to be okay. Hold on to the board. You're going to be okay.'

He paddled to the next girl just a matter of metres away. She was also face down. Gary reached over and tried to pull her up. Shit! She was too heavy. He looked around. Shit! If he couldn't get help, he knew he would probably lose both girls. Time dashed by. Maybe minutes, maybe seconds. Gary didn't know; he was too focused on the moment. Finally, there was reason to count. One boat was coming out of the nearby Patterson River.

'Here. Over here! Help! Help!'

He waved his arms and continued shouting. The driver heard the desperation, and changed his course.

'You're going to be okay,' Gary repeated to the girl hanging on to his board.

But there was no response from the other one. She was about Gary's age. All life had gone from her body.

The boat pulled alongside. Gary helped the conscious girl aboard, and turned to see the driver and another person had hauled the lifeless girl onto the deck. He began climbing onto the back of the boat, edging his tired frame upwards. He was nearly there, but the driver hadn't seen him. The throttle opened, the boat leapt forward and Gary tumbled into the water.

Froth engulfed the following seconds. Gary felt a bang. He had hit something, but as he watched the boat move away, his only thoughts were of sliding onto his board and making it back to dry land. He reached and began paddling with his lower legs bent up in the air. Then, he noticed the blood.

He kept paddling, his feet started to ache and his mind began to wander.

He was in a haze by the time he reached the shore. He didn't know how he had made it, nor could he at first understand why there were two ambulance officers kneeling next to him. Then his feet stung his senses, and he remembered.

'Are the kids okay?' he asked.

He heard no answer.

He saw his Nan and Pop standing nearby. There were police officers too. And other people. Strangers.

Gary was placed on a stretcher and driven away.

The hospital gown fitted loosely, but was certainly better suited to the surrounds than a soaking pair of boardshorts.

A doctor assessed the damage: a cut to the outside of the right foot, a slice to the inside of the left. Twelve stitches. Nothing too serious considering what might have happened. If Gary had fallen a different way, or a moment earlier . . .

well, it wasn't worth contemplating. He had been incredibly lucky that he had only brushed the boat's propeller.

After he'd been stitched and bandaged, he was visited by a policeman who warned him: 'There'll be quite a lot of media at your home when you get there. You're going to get a lot of attention.'

The policeman was right. Gary arrived home with Nan and Pop to find television cameramen, press photographers and journalists waiting.

'How do you feel, Gary?' 'Tell us what happened, Gary.' 'Were you scared?' 'How are your feet?' . . .

Gary felt daunted by the pack of hero-hunters. All he wanted to do was curl up on his bed and feel sorry for himself.

It was only after the cameras and tape machines had stopped rolling and the car park next to Gary's home had thinned of vehicles that Nan and Pop told their grandson that one girl had died. Gary knew she was the one he couldn't lift. In all likelihood she was dead before he'd even reached her. He felt hollow. Nan and Pop told him that was natural. In the weeks ahead, they needed to shelter their 'little boy'.

During that time, Gary received letters from people he didn't even know. The theme was always the same: '*You are a brave young man.*' Later on, he was awarded a scholarship to a private school and was presented with the Royal Humane Society's Bronze Medal for bravery. For this, he 'felt like a nerd' as he was forced to wear a brand-new suit that Nan and Pop bought him so he would look his smartest for his trip to Victoria's Government House for the presentation. Eventually Nan took the medal away from him and put it in safe-keeping until the time came when Gary would appreciate what he'd done. As a young teenager he couldn't understand what all the fuss was about. All he had done was 'paddled out and tried to help a couple of people'.

***

Thirty-seven-year-old Gary Squires has a small scar on either foot to remind him of 14 January 1982. A framed medal and certificate on a wall in his home also take him back to the day when three girls playing with inflatable toys were swept into a deep boating channel by an offshore breeze. It happened near the Patterson River at Carrum Beach in Port Phillip Bay, 40 kilometres south-east of Melbourne.

This beach was Gary's childhood playground. Whenever he felt angry or needed some freedom, he walked out of his home, across a car park and onto the sand. He loved all that he found on the other side of the bitumen: the still days when the whole of Port Phillip Bay was like glass; the wild days when spitting whitecaps thundered in; the quiet of a dawn walk; the bustle of a sweltering afternoon; the thrill of cutting across the face of a wave; the seconds of fear after being dumped; the smell of coconut oil . . .

And all within a mere 50 metres of his home.

Gary was raised by his grandparents, Jim and Rhoda. After his parents separated when he was young, he lived with his mother who, tragically, was forced to be as attached to a kidney dialysis machine as she was to her only child. It meant Gary had a somewhat nomadic existence. While his mother spent long periods in hospital, he camped under one relative's roof to the next, until Nan and Pop decided their grandson's swag shouldn't be swung anymore.

Gary has few memories of his earliest years, but he can recall the day he was playing in the backyard when Rhoda called him inside, sat down with him on his bed, and quietly told him: 'Your mother won't be coming home from hospital. She has died.'

Gary was only seven, yet he'd long accepted that his future was in the hands of Nan and Pop, and that their home was his home. With the scent of a sea breeze through the hallway as strong as the waft of bacon and eggs it was always likely that Gary would grow up with zinc on his nose and sand stuck to his toes. He was a nipper at the Carrum Surf Lifesaving Club, at which he was never short of an uncle, auntie, or a cousin or two to adjust his yellow, red and blue striped cap. Although he liked swimming, there were few greater joys than squirming into his wetsuit and riding the waves, or thumping out the kilometres on a hard run.

Nearly everything he did outside of school required a furious heartbeat. Away from the beach, he loved his footy and Little Athletics, and whether he was competing or simply training, his two most ardent supporters were rarely far away. Nan was the family's courier behind the wheel of a Holden HQ station wagon, although there were times when Pop seemed to do much of the driving from the front passenger seat.

In addition to his job as a factory manager, Pop was an attendant at the Melbourne Cricket Ground, and North Melbourne's Festival Hall. Occasionally he took his grandson along to concerts or fight nights, where the mix of sweat, liniment and hope were pounded into the memory. After he turned eighteen, and was in the zone between leaving school and finding a career, Gary worked for a while at Festival Hall. By this stage he was living between water and fire. He'd climbed the ranks to become his surf club's captain and had also joined the local Country Fire Authority (CFA) brigade, a unit consisting entirely of volunteers. When his pager alerted him to a call-out, he could pump himself full of adrenaline by simply reaching the station to put his overalls on. The sprightly journey from home involved a sprint to the end of the street, a dart and a weave across the multi-lane Nepean Highway, and a scurry over railway lines.

Gary so enjoyed turning out to blazes that he contemplated a career as a full-time CFA firefighter, but he was also being influenced by a mate who'd become a police officer. Either way, it seemed this surf lifesaver's future away from the sun and the Speedos was in a uniform. He eventually applied for jobs at both emergency services at about the same time. However, his road to the CFA was blocked by a technicality. At the time of applying he didn't have his heavy vehicle licence, although he had booked in for a test a few days later. But rules were rules, and he was rejected. Within weeks of that disappointment, he was accepted into the Victorian Police Academy. He was just twenty years old.

The reality of his career shocked him almost immediately. In the same month that Gary started his training, two young Melbourne constables, Steven Tynan and Damian Eyre, were shot dead in an ambush after they'd gone to investigate the discovery of an abandoned car. In the months ahead, the infamous Walsh Street Murders would have an impact on Gary's development as a police officer.

After graduating from the Academy, Probationary Constable Squires was stationed in Frankston, within a short distance from Carrum. While here, he was seconded to the witness protection program for the committal hearing into the Walsh Street Murders. It was a side of policing that he'd never expected to be involved with. With little notice, and no information about where he was going, he was sent to remote locations across Victoria, where he and as many as five other officers, including two from Special Operations Group, would live with a witness for about a week. Although they were allowed to make phone calls, they couldn't tell anyone where they were. Gary was basically forced to put his life on hold for six months. He would go away for seven nights, come back for about three nights, and then leave again.

Once he returned to regular policing he moved through a variety of roles in his early years, including general duties, traffic, and transit (trains) where he'd climbed the ranks to Senior Constable. He maintained his links with fire and water, and was also playing club football. None were prouder of his progress than Nan and Pop, who had seen their grandson overcome the hardships of his childhood to mature into a man with a need to help others. They loved him dearly, although Pop, in his old-fashioned manner, never allowed his feelings to flow into words. But in 1991, there was reason to change. It was three days before Gary was due to compete in Ballarat at the Victorian CFA Championships, a time-honoured event that had first been held in the early 1900s. Gary had nominated for the prestigious hydrant race. In a quaint throwback to the olden days, this race meant each competitor engaged in a series of sprints, which included collecting a hydrant and placing it in a water plug before running to the finish.

'You make sure he goes to that competition,' Pop told Nan. 'And he runs as hard as he can.'

But Gary didn't want to leave; Pop was dying from cancer, and had been told by his doctor that he had just a few days to live. He insisted that his grandson, in so many ways his son, make the start line without looking back. Gary eventually heeded Pop's wishes and packed his bag. Before he left, he did the 'hardest thing I've ever had to do' . . . he said goodbye. And Pop replied with the words that he'd thought every day for 23 years.

'I love you.'

On the morning of the race, Gary received the phone call that he knew was coming. Pop was dead. Later, when the starter's gun went off, he ran as hard as he could with the feeling that he wasn't taking the steps alone. Pop was with him. Together they finished second.

Over the next few years Gary moved through life to the beats of his job, his commitments, his pastimes, and inevitably his love. He married long-term girlfriend Fiona in 1992. They built a house near Carrum, and three years later, they carried baby Hannah over the threshold. Her birth had come at an inopportune time for any sporting father . . . footy finals time. Gary, a wingman, was due to play a sudden-death game for Tooradin in the West Gippsland League at 2 o'clock, Saturday afternoon. Hannah was born at midday. Her Dad was there to hear her first cry, but missed the siren that ended his team's season.

While adapting to nappy changes and spoon feeds, Gary needed a much harder head when he transferred from the transit police to general duties at St Kilda, a cosmopolitan Melbourne suburb which opened a new father's eyes to the vulnerability of youth. He saw fresh-faced teenage girls from respectable homes, with respectable jobs, standing on street corners at not-so respectable hours. Over a mere few months they would be reduced to gaunt ghosts, selling their bodies to earn enough money for their next shot of heroin. It was a miserable cycle in a place that also had 'some pretty ordinary crooks that sometimes you had to bang heads with'. Although much has changed now, St Kilda in the mid to late 1990s brought the worst out of some people.

To avoid the possibility of becoming 'too bitter and twisted with life', Gary sought a quiet change after three years at St

Kilda. He and his young family found peace in Daylesford, a rural town renowned for its mineral springs and spas little more than an hour's drive north-west of Melbourne. Despite the calm surrounds, Gary couldn't resist the call of the beach, and when possible he used to travel back the 90 minutes to Port Phillip Bay to do beach patrols for the Half Moon Bay Surf Lifesaving Club, just a short distance from Carrum. He had switched clubs a few years earlier after becoming the state beach sprint champion. Half Moon Bay had a coach, and Gary had ambitions, but knee injuries from football and a newly discovered passion for surfboat racing eventually ended his days of diving for a stick in the sand.

Firefighting also remained a constant in his life, as he joined the CFA's Daylesford Brigade. On one occasion he swapped vehicles and uniforms in a matter of strides, as he was driven from the police station to the fire station, where he donned his yellow overalls and jumped behind the wheel of the truck when there were no other qualified drivers available to handle a call-out. In many ways, Gary was still a boy on an adventure.

However, the responsibilities of adulthood glared at him when his marriage broke down in 2002. By then, Hannah had a baby sister, Ella. Both girls stayed with their mother while their father returned to Melbourne to work in general duties at Cheltenham, near St Kilda.

It was a traumatic period for all involved, but time, family and friends gradually built new foundations. Gary had limited access to his daughters, and generally only saw them every fortnight. Time together was precious. Just how precious became clear on 17 December 2003, the evening that Gary was forced to rush almost 22 years into his past.

It was about 7.30 pm. The day had been a scorcher, pushing 35 degrees Celsius and pulling hundreds from offices and homes to Gary's old front yard, Carrum Beach. Nan had settled in for the evening in her unit, which wasn't too far away from the old family house. She'd just enjoyed a visit from her 'little boy' and his girls. She waved them goodbye as they made their way to the water with Gary's best mate and fellow police officer Andrew Adams, Andrew's wife Sharon, and his children, ten-year-old Jessica and six-year-old Logan. They'd all just had some fish and chips. Now it was time for the parents to let their dinners go down while their kids played. Three-year-old Ella paddled on the shore, while eight-year-old Hannah was allowed more freedom and had waded through waist-high water with Jessica and Logan to reach a sandbar about 40 metres away. They were on the southern side of a breakwater, a rock wall that stretched alongside the Patterson River and into a boating channel. It was in the same general area that had scarred Gary's feet so many years before.

As the parents sat on the sand watching their children, laughter and chatter floated through the breeze and into the fading light. Christmas cheer was everywhere; what a wonderful way to end the day. Unless . . .

Hannah, Jessica and Logan saw the boy and girl first. They were slapping the water, struggling to keep afloat just ten metres further out.

'Are you alright?' shouted Hannah.

'No!'

The two were being pulled north into the boating channel by a strong rip.

'We'll get help,' said Hannah.

Gary didn't notice his daughter turning towards the shore. Instead, he was distracted by a flash out of the corner of his eye. A man was sprinting into the water. Screams chased him.

'Someone's in trouble!'

Gary stood up and saw a group of people gathering on the seaward tip of the breakwater. Arms were raised, heads were turning.

Hannah, Jessica and Logan tried moving back towards the shore, but they couldn't.

'The kids are stuck too!' yelled Sharon.

Gary was already churning through the rising swell towards the end of the breakwater. He could see three people in trouble, but Hannah and the Adams' children weren't among them. They were out of his eyeline. Andrew, on the edge of the water, and about to follow Gary, changed direction and headed for his kids and Hannah.

Gary passed the man he'd originally seen dash from the shore. He kept going, head down, breaths shortening. The crowd grew larger on the breakwater. A woman jumped off the edge and started swimming. She was the mother of the children Hannah had called out to, and the wife of the man who was now in desperate trouble next to his kids. He'd gone out to rescue them, but instead found himself caught in the rip.

Gary reached them when they were close to the river mouth and being drawn further out, some 60 metres away from the shore.

'We're drowning, we're drowning!' screamed the girl.

She was trying to support her father who was frantically 'climbing the ladder'. His arms scratched the air and his head popped above the water briefly before disappearing again. It wouldn't be long before he wouldn't surface at all. Gary grasped the man's hair, the only part of the body he could see. He pulled hard. The man's head rose above the water, and Gary clutched an arm. Within a splash of seconds, he also held the children.

'You're all going to be okay.'

Gary turned to see the man he'd overtaken near the shore was now close by, but any hopes that he could help were lost in exhaustion.

'Okay, just tread water,' shouted Gary. 'Don't panic. Just try to float on your back. You'll be okay.'

As he assessed his options, Gary was shocked by the mother's arrival.

'Stay calm,' he pleaded.

But she couldn't, as she was desperate to reach her family. She grabbed on to Gary, her arms flailing. Her panic spread to the others, and suddenly Gary was in the middle of a knot of fear that pushed him under the water, and no one was letting go. He fought against the madness: hands, legs, arms, slaps, scratches, mouthfuls of water . . . and a rip that mightn't be defeated. Somehow he reached the surface, and gasped for air.

'Please be calm. We will get out of this. It will be okay!'

A scream in the distance diverted his attention. At first it didn't register, but once he saw three children being dragged off the sandbar 50 metres away, he realised who it was: Hannah, Jessica and Logan. Shit! What? How? If? . . . In a 'split second' he decided to stay where he was; the family he was holding together needed him, and he would only stir more hysteria and create a bigger danger if he moved away. Relief engulfed him when he saw Andrew and another man heading out to the deep water of the boat channel where the children now were.

Andrew rescued Logan first, and handed him to the man who had followed him. He next reached Hannah, who was trembling, and screaming for her father, but Gary was too far away to hear, nor could he see that a young boogie board rider had approached Andrew, and was now taking Hannah to the shore. Jessica too was safe after floating close enough to the breakwater to be lifted out by a woman. With that rescue complete, Andrew headed towards his mate and the terror-stricken family that was now in the middle of the river mouth and being pulled further north. Before Andrew reached them he came across another man who'd jumped in to help. The man was so weary from battling the rip that he was barely managing to keep afloat. Andrew pulled him onto a rubber tube that someone had thrown from the breakwater. After pleading not to be left alone, the man hung on to both the tube and Andrew, as he regained some composure. A boat approached, but incredibly the driver refused to help, and kept going. Thankfully another boat came by, and this time the driver knew his responsibility; two more people would soon be out of danger.

But it hadn't stopped for Gary. His legs were tiring, and his body needed all its remaining strength to control the fear, if not terror, that clung to him. Everyone he was with created a risk. There was the family, and the original would-be rescuer who continued to tread water just a few metres away. Gary knew he was too far from either the shore or the end of the breakwater to try taking one person back at a time. His best hope was . . .

A boat.

'Over here, help!' yelled Gary as he watched a shiny new aluminium single cabin vessel coming in from the sea, rolling over the chop.

'Life jackets! We need life jackets!' shouted Gary.

'What?' replied the driver, struggling to hear over the chug of the boat's engine.

'Life jackets!'

The driver responded, but in his haste, he forgot to do one vital thing. His brand-new boat had brand-new life jackets, complete in plastic wrapping. With both hands weighed down by bodies, Gary now faced the problem of trying to unpack two jackets without losing hold of his cargo. While doing this he convinced the mother to swim to the boat, just a couple of metres away. By now the rip had weakened, and the group floated on the tide past the northern edge of the Patterson River. It was now safe enough for another man to swim out. Gary managed to get the jackets out of their wrapping, and click them onto the boy and girl.

'Take the boy,' he said to the new rescuer.

The man who'd originally swum out made his way across to Gary, and offered the little strength he had left. The father, who by now was very weak, was coughing up water, while the girl, although still frightened, composed herself enough to float in Gary's arms as they passed across the northern edge of the breakwater and slowly stroked their way towards the shore until they reached a point where they could stand up. The feeling of sand against the toes had never been more welcoming. Gary took the girl to shore first, and then returned for the other two. After taking the father in, he laid him down and checked his airways. The daughter sobbed nearby with her brother:

'Is Dad going to be okay?'

'Yes, he'll be fine.'

Gary then ran to the mother, who'd been dropped off by the boat driver further along the river. She too was coughing heavily, and shaking, but wasn't in need of urgent attention. Gary hurried back to the shore and swam out to ensure no one else was in trouble. He reached the breakwater, where he saw Andrew, and asked: 'Is everyone out?'

'Yeah, mate.'

Gary climbed up the rocks of the breakwater to Andrew. Hannah came running towards him. They hugged.

'I'm sorry, Dad. I'm sorry.'

'Don't be sorry. You've done nothing wrong. And we're all okay now. It's over now.'

By that time, Ella was back at Nan's.

The wails of sirens announced the arrival of ambulance and police and, while they worked, two off-duty officers stared out to sea and spoke about the jobs they'd just done. The whole incident had taken ten minutes, fifteen at the most, but for those trapped in the middle, it had seemed so much longer. While the rip had pulled, time had dragged, and in doing so, one man was taken back to his childhood. Gary Squires had twice been at the right place at the right time on the same beach nearly 22 years apart. Nowadays, he shrugs when reflecting on both incidents, and with an endearing modesty he simply says: 'I was there, something had to be done, and I did it. We all did.'

As opposed to a commonly held view, heroics don't necessarily lead to euphoria. A few days after the second incident, Gary received a phone call from Andrew:

'How do you feel?' asked the father of Jessica and Logan.

'Shithouse. What about you?'

'I'm a mess.'

'Well, that's how I feel too.'

The mates arranged to return to the scene of the rescues and have a swim to convince themselves that Carrum Beach was a safe place. They also talked about the incident over a few beers at a local pub. It was important that they shared their thoughts and feelings, and supported each other. In the following weeks they were counselled by a police psychologist who warned them that they 'might be on a roller-coaster with

Justin Morrison and his wife, Sarah, with Georgia in January 2000 at Lilli Pilli on the south coast of New South Wales.

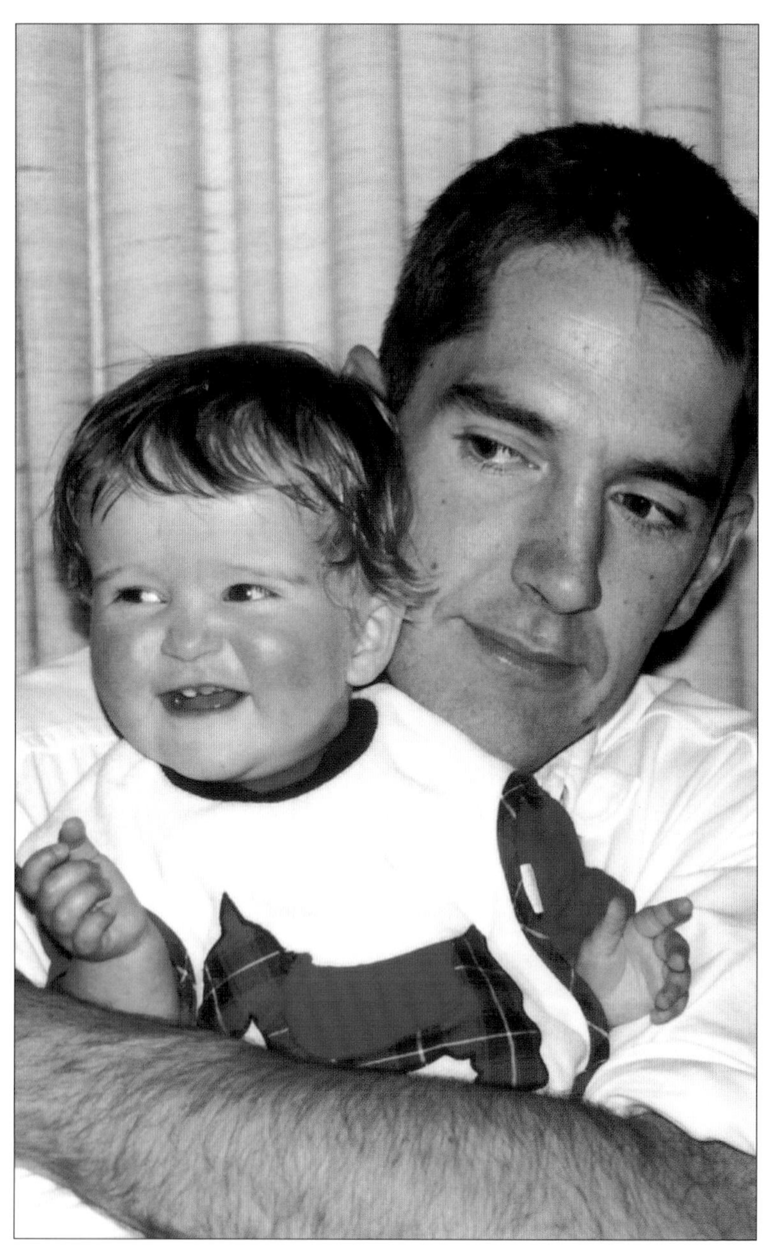

Justin and Georgia. 'Parents shouldn't have to go through that …'

The Pitman family (*standing, left to right*) Superintendent Grant Pitman, Inspector Stacey Pitman, Superintendent Garth Pitman, Inspector Kendall Pitman. Brian Pitman (*seated*).

The barefooted Ivy Rooks and her husband, Tjamme, on their prized Yamaha V-Stars.

Rescue Paramedic Neale Smith at home in the Tasmanian wilderness.

Detective Superintendent Deb Wallace. 'A westie through and through … and proud of it.'

Paramedic Dave Stevenson and his children, Rescue Paramedic Pat and Intensive Care Paramedic Kerrie. 'Ambulance is family.'

Thirteen-year-old Gary Squires with his surfboard and bandaged feet after his brave rescue of three young girls at Carrum Beach in January 1982. Pop is standing behind him. (Courtesy *Sun* newspaper.)

Gary accepting the Royal Humane Society's Bronze Medal for bravery from Victorian Governor Brian Murray.

Senior Constable Gary Squires with his daughters Ella (*left*) and Hannah at Carrum Beach. (Courtesy *PoliceLife* magazine.)

The Avoca Angels (*standing, left to right*) Shirley Squires, Margaret Dennis, Sandi-Lee Squires, Jackie Squires. (*Middle row, left to right*) Mary Knowles, Helen Johnson. (*Front row*) Helen Reynolds. 'We've never had a fight or argument in the group.'

emotions'. That's exactly what happened to Gary, as he posed the 'what ifs': What if he hadn't been there? What if he had been pushed under the water and hadn't been able to surface again? What if no one had seen Hannah and the Adams' children? What if, what if, what if . . . And the most torturous question of all: What if Hannah had died? The incident had 'given him a different spin on parenthood', and a greater understanding of what others endure at times of grief. It was something that he was too young to fully understand when he dashed into troubled waters as a thirteen-year-old, but as an adult and a police officer, he'd come to realise how brittle life was, and how a single, often unexpected moment, could change the lives of many forever.

Eight months after his second rescue at Carrum Beach, Gary received the Bravery Medal, which is presented through the National Honours List and acknowledges 'acts of bravery in hazardous circumstances' by either members of the military or civilians. This time there was no feeling like a nerd as his police tunic sat comfortably across his broad shoulders during the ceremony at Victoria's Government House. Andrew was also there, receiving the Commendation for Brave Conduct.

Medals won't change Gary. He's a typical Aussie bloke who loves his family, great mates, a few beers and a good time. These days, he still takes the board out when he can, but he no longer has the time to volunteer for the CFA. He visits Nan regularly, and Nan still calls him her 'little boy'. He has married again after falling in love with Lara, a fellow surfboat rower whom he met at a 'boaties convention' on the Gold Coast. Fittingly, he is now a Senior Constable with the Water Police, working in an office that is much bigger than the front yard of his childhood. That front yard has proved to be a metaphor for his life. Gary knows only too well what ebbs and flows can bring. Hannah and Ella, too, are learning this as Nippers at Half Moon Bay. As they discover for themselves the Squires' passion for water, their father has already made the decision that when he dies his ashes will be spread near the mouth of the Patterson River.

IVY ROOKS

'Toe Rings and Full Throttles'

'There is that moment when you're almost overwhelmed and start thinking: "Shit! What am I doing this for?"'

Volunteer Ambulance Officer Ivy Rooks, Tasmanian Ambulance Service

She rides a Yamaha 1100 V-Star motorcycle. She has more toe rings than shoes, and thinks nothing of walking over frost in bare feet. As a girl she ate kangaroo patties, rabbit pies, and mutton birds; as an adult she cooks an irresistible meatball soup. She had her first child when sixteen, was married at eighteen, and now, approaching 50, is proudly a grandmother. She can sing country rock and can carry a tune on the bass guitar. She lives in a cottage with her husband, Tjamme (pronounced Chum), her two cats, Gizmo and Smudge, and her two dogs Max and Shylo. The woman I am describing is Ivy Rooks, and she really knows what it means to say 'Oh Shit!'

The 'Oh Shit Factor' occurred late on Sunday 18 February 2001, but before we visit that dramatic night, let's wind back to an unusual childhood filled with dance halls, button accordions and bonnet surfing.

Montagu is a small rural settlement in Tasmania's far north-west. It was home to Fred Zeuschner, who battled to make a living on a dairy farm. The days were full of labouring, milking and feeding, three fundamentals of cattle farming that developed new meaning the day Fred's pregnant wife Audrey felt her contractions couldn't be ignored any longer. She sent her husband hurrying away to find a doctor, but it didn't help that his fastest mode of transport was a tractor. By the time he returned, Audrey was cradling a healthy baby girl.

Ivy Zeuschner was the first of ten children, two of which died in infancy. She remembers having 'kids around me all the time', and the older she became the more responsibility she assumed, helping her mother 'who always seemed to be pregnant'.

The forever-increasing Zeuschner family moved almost as frequently as Audrey felt contractions. They seemed caught in a cycle of life in which each new child added to the pressure on Fred to support his family, and so he went wherever he could to find steady employment: the nearby Robbins Island as a farm contractor; the town of Burnie and the villages of Howth and Fingal as a railway labourer; and, surprisingly, considering his affinity with small settlements, he even travelled as far as Melbourne to work in a factory. The pace and size of the Victorian capital was too much for Ivy, who was then in her early teens. She describes it as 'the worst eleven months of my life'. Hustle and bustle hadn't been, and wouldn't be, part of her life. She loved simplicity, which was best underlined by her preference for bare feet over shoes. It wasn't any type of statement other than to say 'my toes are more comfortable this way'. Although she obeyed regulations at school, whenever she had the chance to slip off her shoes and socks, she hastily did so. Only once was she embarrassed by this, when she was admonished by an old man while walking along a street.

'Can't your Dad afford to buy you a new pair of shoes?' he said scornfully.

It so upset Ivy that she thought of running home and plunging into conformity, but she soon shrugged the incident off, assuring herself, 'Well, my feet are healthier than his!'

Whether in bare feet or shoes, Ivy wandered and occasionally waltzed through her childhood. She relished the times when the whole family went to dance nights at the numerous local halls. Fred played both the button accordion and the guitar, and as he struck the notes, his children slipped and slid across the floorboards, cheerfully singing along. As an evening wore on, the younger ones would curl up on benches tucked against walls, or wearily make their way to the softer comfort of the bench seats in the family car. There was also the option of the warm cradle made by their big sister's arms—but you had to be quick to be the first.

'I was always holding someone,' Ivy recalls, 'or stopping someone from doing something they shouldn't. And I was always putting a Bandaid on someone. Mum was hopeless with any accident, no matter how small it was. She'd panic, so I was always the one that dealt with it in a calm way when someone fell over or burnt themselves or did whatever. Mum would tell me what to do, then I would do it.'

Ivy was destined to enter an adult's world sooner rather than later. She left school when only fifteen, at which age she was again on the move with her family to King Island, in Bass Strait. Buffeted by the notorious winds of the Roaring Forties, the island was renowned for its shipwrecks, but at the time of the Zeuschner family's arrival, maritime infamy had surrendered to agricultural fame, particularly dairy produce. It was here, among a sparse scattering of 2000 people, that the Zeuschner clan found stability. Fred worked in the abattoirs near the island's commercial hub, Currie, and although underage, Ivy poured beers and waited on tables at a club.

In such places as Currie, people are forced to make their own fun. Ivy sought hers in fishing, barbecues and picnics, and the precarious pastime of riding on old car bonnets tied to the back of vehicles, which churned through sand dunes before dashing over the harder beach surfaces. With a quick turn of the wheel and a squeeze of the brakes, a driver could send a bonnet-surfer bouncing into the water and across the waves. It was an activity full of white knuckles, windswept hair, and shrieks of nervous laughter.

Such joy-riding took an unexpected turn when Ivy met the driver of a battered EH Holden that had no doors. It had been bought for twenty dollars by Tjamme, a steely-eyed council worker of Dutch descent whose parents raised twelve children, joking that it was cheaper by the dozen. Their friendship began with a date at the local hall to watch 'some crappy vampire movie', but it soon grew to more than mateship. They best found time alone by sneaking out of town at night in the doorless sedan.

The relationship grew further when Ivy gave birth to a boy, Jamie. She married Tjamme two years later, and within another four years they had added two daughters, Sarah and Sacha, to the family. With neither parent wanting to uphold their family traditions of having more children than bedrooms, three was considered enough of a crowd. However, one tradition remained—the need to keep moving in search of better opportunities. After quitting his council job, Tjamme worked in a scheelite mine, and when he wasn't in the mill, he helped on his father's dairy share farm. But by 1980, after twenty years on the island, he and Ivy packed up the family and headed for Tasmania.

After a few short stops, their travels led them to Waratah, a town in the north-west wilderness that once boasted one of the largest tin mines in the world. It was an isolated spot; the nearest main town, Burnie, was more than an hour's drive away, along a road that wound through the treacherous Hellyer Gorge. Apart from hearing the rumble of logging trucks, and the occasional swarm of camper vans, buses, and rental cars during the warmer months, Waratah was a quiet spot where locals passed the hours at their own pace. Although it had few major amenities, it appealed to Ivy. After having one hand on a suitcase for so many of her 23 years, she had a feeling she had come to a place which wouldn't be waving her goodbye in a hurry. She was at home. There was no hustle, no bustle; just a simple approach to living day by day.

They quickly settled into the pattern of small town life. While Tjamme worked at the new Que River Silver Lead and Zinc Mine, Ivy raised the children. By the time Jamie, Sarah and Sacha were all at school Mum, too, entered the classroom as a volunteer teacher's aide. She soon graduated to a paid job as gardener, cleaner and special teacher aide. Whether she was pulling out weeds or sweeping corridors, Ivy had a readily identifiable trademark—bare feet.

In addition to becoming increasingly involved in school events as both an employee and a mother, Ivy, eager to help the whole Waratah community, joined the town's hospital auxiliary. There was no hospital as such, just a small medical centre that was staffed by a sister and an ambulance driver who had no first aid qualifications. Initially Ivy didn't think she would serve any greater purpose than being a fund-raiser and part-time receptionist, but her outlook changed when the auxiliary received a letter from the Tasmanian Ambulance Service seeking volunteer officers.

More than half of Tasmania's population lives outside its capital city in small, often isolated places. This makes it difficult to provide professional health services to all areas and, as a result, volunteer officers are considered essential. With a strong sense of community values, and a need to help others that was fostered by her childhood, Ivy decided it was time to contribute more.

Together with a few other auxiliary members, she enrolled in a two-week training course that was held at Savage River, a town 40 kilometres away. Each day they piled into a van that had only two front seats, and an empty rear compartment that was filled with bean bags to soften any bump along the winding road to and from classes. After learning about everything from cell formation and the anatomy of the heart, to basic patient care and cardiopulmonary resuscitation (CPR), Ivy and her fellow students graduated as volunteer ambulance officers.

Apart from receiving a certificate, each new worker was given a white shirt with the ambulance emblem, a red cross against a green Tasmania sewn on the sleeve. The rest of their uniform was their choice. Trousers and shoes were the obvious additions needed, but this created a problem for Ivy who didn't have any adequate footwear. She went and bought a pair of boots, after which her penchant for having free-and-easy toes when working was only noticeable when she was in the ambulance itself, or attending training sessions. The sight of her bare feet on such occasions forged a reputation that exists to this day. When an officer mentions to a colleague that he has seen Ivy, the follow-up question is commonly: 'Was she wearing shoes?'

Ivy's 'career' in the ambulance service began in 1984. The most distressing of her early jobs was attending to a baby boy who'd died from Sudden Infant Death Syndrome (SIDS). She also had to deal with being the first on the scene at a fatal car accident. However, most of her experiences didn't have the force to play on her emotions. Although her chief responsibility was to respond to pages that were sent by the ambulance service's central communications point (Comms), it didn't stop her from receiving phone calls at home, or knocks on the door asking for help. Amid the treatment of cuts and bruises, sneezes and wheezes, she offered advice to some, simply listened as a companion to others, and occasionally pulled stitches out of dogs and cats. There were also the serious yet comical moments, such as the time she treated a miner who couldn't feel his genitals after he'd become wedged at the hips in a piece of machinery. If the concern of losing his manhood wasn't frightening enough, the prospect of facing his wife, who was eager to bear children, most certainly was. Upon inspection, Ivy discovered his testicles had been forced partially into the abdomen, but with a few well-placed fingers and a bit of pressure, she managed to return the absentees to their rightful place, much to the relief of the obviously anxious patient.

Then there was the case involving a man complaining of back pain after a car accident. As part of the treatment, Ivy and her partner, Janny, who also happened to be Tjamme's sister, needed to 'expose', or undress the man before immobilising him on a spinal board. One pair of jeans, two pairs of long thermals, and two pairs of boxer shorts later, the patient was finally down to his underwear. The whole process had taken some considerable time and effort, not least because the man was a Japanese tourist who barely understood a word of English, and was stunned to find two Australian women so eager to take his clothes off.

There was also the memorable day when a woman with a fractured ankle was struck by an urgent call of nature while travelling in an ambulance being driven by Ivy. She refused to use a bedpan, but thankfully a toilet block was close by, so Ivy drove her vehicle as close to the amenity as possible. With the patient in little pain, but reluctant to hop the short distance remaining, she was wheeled on a stretcher into the disabled toilet. From there, she slid off the stretcher's edge and hopped to safety just in time. Then the whole process was reversed while all three struggled to control their laughter.

As with so many emergency service workers across the world, Ivy feels humour is an important, if not essential companion to have close by at any job. Sometimes it's needed on the way to an incident, sometimes it's needed at the scene, sometimes it's needed afterwards, but sometimes . . .

It was 10.30 pm, Sunday 18 February 2001. Ivy was at home preparing to go to bed when her pager beeped. A minibus had overturned on Dove Lake Road near Cradle Mountain, one of Tasmania's main tourist attractions, about 40 minutes' drive south-east of Waratah. Initial reports stated there were three to five injured passengers, but there were no firm details. Ivy followed her standard procedure. She phoned the communications centre to confirm she had received the page and was on her way, as were numerous other volunteer and professional units in the area, including paramedics from Burnie. As was often the case with an officer on call, Ivy had the ambulance parked at home. Before leaving Waratah, she picked up two other volunteers: Maria, a qualified officer; and Derek, who had only finished his assessment course earlier that day. This was to be his first job.

Ivy told her colleagues to 'buckle up and hang on because this is going to be a quick trip'. Officially volunteers have to adhere to the speed limits, even when driving under lights and siren, but with dry conditions and little traffic on the road, unofficially Ivy put her foot down. A shod foot.

As they drove, they discussed a variety of scenarios they might find on arrival. They'd been taught how to handle the very worst accident scenes, but that was only in training; reality was always a bit harder than imagination.

They turned into Dove Lake Road, a narrow, unsealed stretch flanked by precarious slopes that plummeted and climbed into darkness. Approaching each bend and hill, Ivy wondered what she would find on the other side. She'd travelled this road many times before, but had never needed to shock the bush and corrugated dirt with emergency red. As the ambulance lights continued flashing, Ivy turned another corner, and slowed. She looked ahead, and saw the 'Oh Shit Factor' staring back at her.

There were people standing, wandering, sitting, crouching, lying on the road. Some park rangers' vehicles were pulled off to the side, while deep below them, swords of torchlight struck each other in silence around an upturned minibus. A ranger ran to the ambulance and informed the crew what had happened. The bus had rolled over the road's edge and had come to rest on its roof 30 metres down a 45-degree slope. Eighteen people had been on board. At least three were dead. Ivy recalls:

'There is that moment when you're almost overwhelmed and start thinking: "Shit! What am I doing this for?" But then you take a deep breath and say DRABC [Danger, Response, Airway, Breathing, Circulation]. It's what it comes down to. It doesn't matter whether you've cut your fingers on a little knife or a sawblade, it still comes down to the same basics. If you go back to your training, the "Oh Shit Factor" will sit back a bit.'

Ivy radioed central communications, and gave the simple but telling request: 'Send us everything you've got!

By this stage, communications had already received updated information about the size and seriousness of the situation. Further units had been dispatched, including a rescue helicopter.

As Ivy further assessed the scene she was taken by how quiet it was. There was no screaming, no yelling, no hysteria. It was 'like watching a movie with the sound down'.

The crew's initial job was to find all eighteen accident victims, a task made challenging by the presence of dozens of other tourists. Five patients were identified wandering on the road. None was in a critical condition.

Ivy and the crew then headed down the slope, pushing through metre-high undergrowth with only hand-held torches to light their way. The next person was found about halfway down. Ivy checked his carotid pulse, turned her back and moved on. Her actions seemed harsh, but this was no time to treat the dead.

One by one, the passengers were found. Two were dead under the bus, and another, a woman, was crying for help on the slope below the crumpled vehicle. Her admission to Ivy was chilling: 'I'm a nurse. I know I'm dying.'

The woman had extensive internal injuries and pressure was building across her chest. Ivy gave her oxygen and tried to comfort her, but there was nothing else that could be done. Unless she underwent emergency surgery, her chances of survival were remote. Time was against her.

The minutes dashed by. A distressed husband asked Ivy if she'd seen his wife; bystanders asked how they could help; central communications asked for information; and Ivy asked herself how best to handle it all. She appointed people to keep the injured company, leaving each with the direction: 'If you have any problems, call for Ivy.'

About twenty minutes passed before another volunteer ambulance crew arrived. Soon afterwards the professional crew arrived from Burnie, and as the officer first on scene, Ivy officially handed over control to her senior colleagues. But her work wasn't done, she and other officers began putting patients on spinal boards and carting them to the roadside, where they were safely unloaded and treated.

The patients ranged from needing only basic first aid to urgent hospitalisation. The nurse required most attention. She survived long enough for the helicopter to arrive, but died as she was being loaded in.

'She knew what was happening to her all the way up until she died,' recalls Ivy. 'It was the saddest part of the whole night. Whenever cases like that happen I always try to think of what an instructor told me during my initial training: "When a person dies, you must tell yourself that you were the best chance that person had. You can't always do everything you want to, but you do the best you can, and that means the patient had the best chance they could have had at that moment."'

Nearly ten hours after receiving the call-out, Ivy was home again. She'd first gone to the ambulance centre at Burnie, where all officers were asked if they wanted counselling. Ivy declined.

She had 'never had a real problem with death, and never got pent-up or too disturbed'. She dealt with it by simply moving on. After having a shower and freshening up, she went to work at school. Tjamme, who was on a mine shift, called her once he'd heard what had happened.

In the following days, the details of the accident became clearer. The bus group had been on a spotlight excursion for wildlife. Just half an hour into the trip, the driver pulled off to the side of the road to allow an oncoming vehicle to pass, but the road's edges crumbled, and the bus tumbled down the slope. Sixteen of the eighteen people were tourists, nine from overseas.

While the surviving passengers found their own ways to put the accident behind them, Ivy returned to her life in Waratah. She spent a short period as an ambulance officer at the Savage River Mine, and nowadays works at a call-centre in Burnie, but her pager is never far away. To this day, she still occasionally thinks of the 'Oh Shit' night, but has never 'felt her mind play tricks'.

Ivy does admit, 'I do get emotional talking about it, but it has never gone around in my mind all the time. Some people have told me it does catch up with you after a while. You might be fine for years, then one strange little job might send you off, but hopefully that won't ever happen to me. Sometimes I wonder if I could have somehow made a bigger difference. You're only human to do that, but those thoughts don't stay for long. It is the most dramatic experience I've ever had, and I hope I never have another one, but if I do, I know I will be prepared.'

Although she has never sought recognition for what she does, Ivy was touched when she received a thank you letter from a woman involved in the Cradle Mountain accident. It was a simple gesture, but one that will never be forgotten. Occasionally Ivy receives other letters and cards, mostly from people on the mainland who express their surprise ambulance officers can be volunteers. Over the years, Ivy has received numerous honours, including the National Service Medal, Ambulance Service Medal in the Queen's Birthday Honours List, and in 2000 she 'shuffled along like Cliff Young' near Burnie as a member of the Sydney Olympics Torch Relay.

However, amid the gains, there have also been losses. Both her parents have died, and all of her children have moved away from Waratah. Though it is hard to let your children go, this has allowed Ivy more time to indulge in her hobbies, including photography and furniture-making. But nothing thrills her more than opening up the throttle of her Yamaha V-Star, with Tjamme rumbling along behind her on a twin bike. At the time of writing they are planning a road trip to Western Australia. Unlike so many other journeys Ivy has been on during her life, this time she won't be looking for a new home. A tiny cottage in a remote settlement in Tasmania's north-west wilderness is all she needs. Waratah is lucky it has such a woman. Bare feet and all.

JUSTIN MORRISON

'Such is Life'

'Death is very much a part of life.'

First Class Firefighter Justin Morrison, ACT Fire Brigade

Tony Morrison was a bushman. His face had been worn by the seasons, both good and bad. He was a shrewd bloke who reckoned he'd seen just about all there was to see in life. He had felt it too, especially in his back which he'd broken while mustering stock in his twenties; osteoporosis and the years had since combined cruelly to mould a stoop.

Although his weathered exterior underlined a toughness that comes with country life, Tony also had a sentimental side. He was a renowned bowerbird, hoarding bits and pieces of memories; old farm records were among his favourites, including a contract signed with an X to mark his grandmother's signature. He also relished reminiscing. With a rolled cigarette never far from his fingertips, he recalled the days when he and other stock owners in the area drove sheep to the grassy plains near

Kiandra in the Snowy Mountains. The springtime trips took a couple of days, but the hours passed quickly. This was due in no small part to the effects of the cargo carried by the reliable packhorse 'Carry Grog'. Whether or not the story was embellished through the years by a bushman's sense of humour, no one argued with Tony's tale that his lager-laden labourer was so smart that it could judge the gap between trees, knowing when it was safe to pass through the space, or when a gentle change of gait and a swerve were needed. Back then 'Carry Grog' wasn't the only animal of burden; this was the era when the sheep's back carried Australia, and it seemed no load was too great for the bush to lug.

Tony had initially thought of pursuing a legal career, but being the only son in a family of three, he was handed down the property 'Rob Roy', a sheep station in the Royalla district just a handful of rolling hills outside Canberra. It had been in the Morrison family for four generations. By the time Tony married Catherine, a devout Catholic woman from Sydney, the bush was his future. They raised a daughter and four sons, the youngest of whom, Justin, was born in 1967.

Justin's childhood was typical of many country children. There were the common themes: machinery to climb; poddy lambs to feed; motorbikes to ride; logs to look under; rifles to fire (but only once he turned eleven, and only after being drummed with repetitive safety drills by his father). There was also the less conventional pastime of rearing 'Wally', a baby wallaby whose mother was killed on the road. As the little buck grew stronger, he enjoyed ambushing the family's blue heeler dog by jumping out from under the kitchen table with claws at the ready.

It was a childhood filled with family love and country heart, but amid the joys there were also hardships, and none signalled more alarm than clouds of smoke and the cracks of exploding eucalypts. Justin's first experience of a bushfire came when he was about five. He recalls his father leaving home to fight the flames in the Tinderry Ranges between Canberra and Cooma. Tony returned a few days later and announced: 'It's gone. They'll stop it at the coast.'

Over the years Tony had been a respected member of numerous bushfire brigades: Tuggeranong, Tharwa, Williamsdale and Guises Creek. His children joined him once they were considered old enough to thump a wet hessian bag into flames without getting into trouble. The rules of the brigades were simple: look after yourself and your neighbours.

'If a fire started on old Joe Bloggs's place, it was drop tools, stop whatever you were doing, and away you go,' recalls Justin. 'Back then each farm pretty well had its own truck, a pump and a water tank. We got an old army dump truck that took Dad a couple of days to drive back from a depot in Sydney. It was an International with a crash gearbox. We all got together and painted it red, and during summer left the tank full of water ready to go.'

A fire didn't just mean work, they also were social occasions. As the trucks grunted along the Monaro Highway (the easiest access to most properties), people would wait by the roadside to be picked up. They'd climb on and cling to the sides, or cram into the cabin. Everyone had done enough training to know to 'put the wet stuff on the hot stuff', and most of the time this practice returned the desired results. Once a fire was beaten, the fighters would lean against their trucks in the scorched paddocks and eat sandwiches made by their wives, mothers, daughters and sisters. With some cartons of beer to soften the crusts, talk soon enough turned to how big the flames were, and how they were tamed. These weren't the only social gatherings fuelled by fire; there were dances, barbecues, and field days, all of which raised funds for the purchase of equipment. Fire may be destructive but in Royalla it could also build closer relationships. Nevertheless, it remained Tony's biggest fear. The prospect of seeing his livelihood in cinders was never far from his thoughts during summer. His property had a mixture of flat paddocks and others that climbed into the hills, where clumps of eucalypts kept watch over the district. If the conditions were right, no amount of community spirit could dampen a raging front.

Such worries didn't dominate Justin's thoughts. By 1985, he'd grown into a tall, lean young man with other concerns, chiefly his final year at St Edmund's College, where he wanted to earn enough marks in his exams to gain a place doing Agricultural Economics at university. His father had stressed to all his children that they needed to pursue interests away from 'Rob Roy'. Just as one season can turn its back on the next, times on the land had changed; the romance of the 'pound for pound' days seemed almost forgotten in the shadows of the modern world's vast and complicated issues.

One March afternoon Justin walked out of class, bag over his shoulder, folder in his arms. It was 3.30 pm. Nothing seemed unusual until he reached the bus stop and noticed a pall of smoke hanging in the sky in the direction of Royalla. He immediately thought: 'That's out near home.'

Wherever the fire was, the damage could be great. The grass was waist-high and dry, the temperature was well into the thirties, and the wind was a howling westerly. All Justin could do was sit on the bus and peer through the window at the thickening clouds of grey. The trip to 'Rob Roy' usually took about half an hour, but not on this day; the bus was stopped by police roadblocks just before the Monaro Highway wound up into the Royalla Hills. Justin felt helpless; he still didn't know if the Morrison property was in danger, but the sight of the hills burning with a roar that sounded like 'four or five jumbos taking off' made him think the worst.

He waited and hoped.

Meanwhile, two of his brothers, Grahame and Gerard, were fighting a smaller fire in the nearby Queanbeyan district when they received the news: 'There's a fire near "Rob Roy".' They hurried off, but by the time they arrived home, the flames were in command, firstly racing across the flats of a neighbouring property, then into the hills at the back of their own place. With the westerly whipping the front, the fire galloped through the grass, and leapt from tree to tree. It was unstoppable.

It was dark when the roadblocks were finally dismantled. After needing a lift from a relative to get to 'Rob Roy', Justin recalls staring out at the devastation.

'It was pitch-black. There were embers glowing everywhere. Logs and fence posts were still burning, and the smell was so strong that you could taste it. It was quite surreal.'

By then, the fire was under control. In just a matter of hours its front had cut a path through 'Rob Roy' and two neighbouring properties. As he assessed the ruins the following morning, Tony was thankful for his practice of keeping the paddocks closest to the house and woolshed eaten down. The land was 'black as far as the eye could see', except for a brown square where the fire had run out of fuel around the buildings. However, nothing could save the sheep. In the days afterwards, Justin joined his father and brothers for one of the most distressing jobs a grazier can ever do . . . shooting stock. It was heart-breaking to see the animals in agony. Some had melted ears, and many tried to push themselves along on their knuckles because their coronets had all but disappeared. Fifteen hundred sheep had already died or were waiting for the bullet. Once killed, they were buried in a pit.

It was a traumatic time for the Morrisons. Amid the despair there was also anger when it was discovered that the fire had been deliberately lit by a person passing through the area on a motorbike. The offender should be grateful that he was never caught by a Royalla grazier.

'The incident affected us all,' says Justin. 'For me, I got a greater appreciation of what fire could do. I had seen what could happen at smaller fires, bonfires, and burn-offs, but I hadn't been exposed to anyone or anything being injured. Seeing the sheep and other animals wounded gave me a greater respect for fire. Pretty much anything at a given temperature is going to burn or change state. I hadn't realised that until then.'

Slowly, 'Rob Roy' recovered and as its fences again began dividing the land and new mobs of sheep dotted the paddocks, Justin did well enough at school to be accepted into his preferred course at the University of New England in Armidale, New South Wales. It wasn't as he'd expected, and he soon transferred to study an arts degree with a major in economics and geography. It was here he met Sarah, an outgoing student who complemented Justin's quieter nature. They shared a house together in a purely platonic relationship, neither knowing what would happen years after they'd left tertiary education behind.

After graduating, Justin returned to 'Rob Roy' without a clear career path. He learnt to shear, and laboured on other properties to supplement his income when wool went through its troughs. By the early 90s he was also working as a Qantas baggage handler at Canberra airport.

While his professional future lacked some direction, another part of his life was blooming. As an assistant to Federal Senator

David Brownhill from New South Wales, Sarah made regular trips to Canberra. What began as social phone calls to Justin—'I'm coming down this week, do you want to catch up?'—soon stepped beyond a friendship. History had played a similar number more than 30 years earlier when Tony met Catherine, a parliamentary staff member, who at one stage worked for Harold Holt, a future Prime Minister.

Justin and Sarah moved in together in Canberra, a situation that prompted Justin to seriously re-assess his work options. A sales job in a liquor store gave him more income but little stability. In his search for security his interest grew in what his third-oldest brother was doing. After being overlooked four years earlier, Gerard was reapplying to become a firefighter with the ACT Fire Brigade. There was no definitive moment that swayed Justin to do likewise, but after considerable thought he decided to 'give it a crack'. After all, it would provide him with a regular pay cheque and a career path. It was 1996. Several months later, Gerard received a letter of acceptance but a different envelope carried bad news for Justin. At a time when Grahame was working in the construction industry, Paul was pursuing a banking career, Margot was married and living and working in Sydney, and Gerard was starting afresh, the youngest Morrison child was still searching.

He returned to Qantas, and worked in a variety of roles including check-ins and movement control on the tarmac. On 25 October 1997 he married Sarah at St Gregory's, Queanbeyan, the same church in which his sister, grandmother, and great-grandmother all walked down the aisle. While upholding such family traditions was cause for a few additional laughs and back-slaps, for Justin and Sarah nothing could compare to the event that occured on 16 November the following year, the day Tony turned 64. But that wasn't the reason for celebrations.

The real joy came at Canberra's John James Memorial Hospital when Georgia Ruby Morrison was born. Justin recalls:

'I remember walking out of hospital proud as punch, thinking "Isn't life good!". I'm going to show Georgia a good life. I'm going to do things for her that I've never ever done myself. I'm going to be the best father I can be.'

Sarah, too, was floating. Lying in her hospital bed, she sipped chardonnay, ate chocolates, and watched *Steel Magnolias*. The significance of the day was overwhelming. Apart from sharing a birthday with Tony, Georgia had arrived in the middle of shearing week at 'Rob Roy'; perhaps her timing could have been better, but Grandpa Tone thought anyone born during such an important time was destined to be very special.

Four days later, the Morrisons' new world changed. After having problems feeding, Georgia was transferred to the Royal Children's Hospital in Melbourne where she was diagnosed with Truncus Arteriosus, a complex congenital heart defect which affects 1 in 10 000 babies. Instead of having a pulmonary artery and an aorta, she had just one main blood vessel leaving the heart. Georgia Ruby Morrison was indeed a very special person. At just nine weeks old she underwent open-heart surgery. Cold statistics suggested she had a good chance of recovery, but that was only the beginning. She spent seven months in four different hospitals. Cardiology Ward 7 West at the Royal Children's Hospital was her home: the electronic blips on monitors were a long way from the bleats of the newborn lambs at 'Rob Roy'. Justin and Sarah moved into Ronald McDonald House, a support base for the relatives of seriously ill children. Every day they hoped against hope that their daughter would recover.

At Christmas, Justin was temporarily reminded of his failed attempt at a new career when firefighters from Melbourne's

Metropolitan Fire and Emergency Services Board carried in loads of toys and cheer into the Children's Hospital.

'They came around with a big brass band and handed out soft toys to the kids. I thought, "How good are these blokes to do this!" It brought a tear to my eye to see all the fire trucks outside and all the kids smiling. The whole spirit of it all touched me.'

Georgia was eventually allowed to go back to Canberra. She was baptised at home, and over the following months she showed that her heart may have been weak, but it also gave her strength. Family, friends and medical staff agreed that her eyes told of an 'old soul'. They were fitting words, for few people are forced to endure in a lifetime what Georgia did. At nine months, she had another open-heart operation. The fight continued as the list of complications grew: double valve incompetence, parvovirus, upper respiratory virus, rotavirus, pyelonephritis, and eventually cardiomyopathy, a disease of the heart muscle. To an outsider these are just meaningless terms, but to Justin and Sarah each further tore the fabric of a 'world that was falling apart'.

In many ways Georgia was like other children her age. She loved music, and although her illness restricted her movements, she still managed to move her head in a bouncing rhythm to her favourite songs. In between the frequent flights to Melbourne for treatment, she visited 'Rob Roy' and 'Deasey's Cutting', the farm owned by Sarah's parents. She loved the outdoors; it was in the Morrison blood. In January 2000, she was taken to the beach for the first time at Lilli Pilli on the south coast of New South Wales. She laughed, she touched, she listened, she talked, she watched . . .

Then, on 17 February Georgia died. She was at home with Justin, Sarah and Catherine.

'I don't talk about it unless I've been asked,' says Justin. 'Parents shouldn't have to go through that. In the end we just had to watch her die. The amount of tests, all she went through, were too much. The only other option was a heart transplant, but we took her home. It wasn't fair to do any more to her. It took three to four hours for her body to start shutting down. We piled all the love we could on her, then sent her on her way, hoping that the place called heaven did exist. If so, there were plenty of other relations up there waiting to greet her. Afterwards, there was an enormous emptiness in our lives. We'd spent so much time just living for her. All our friends were pretty well married at the same time, and having kids at the same time. It affected them, it affected us. It was new to us all. Mum and Dad weren't the same from that day on. It sort of mortally wounded them, I think. My old man lost a bit of spark and spirit. You could see it in his eyes. Mum too, but she was stoic and didn't really let us know how she felt.'

In the weeks and months that followed, Justin sought an escape through his work at Qantas, but he needed more, and again applied to become a firefighter. The first step was a written application outlining communications skills, physical fitness, ability to be a team member, work history and relevant qualifications (the minimum requirement was a medium rigid truck licence and a senior first aid certificate). This time the selection panel was interested, so Justin advanced to an extensive exam stage that included maths, word associations, aptitude tests, and spatial reasoning. There was also a medical, and an exhaustive list of physical capability trials: carrying a 60-kilogram dummy over a short distance; walking along narrow beams; holding hoses; finding your way out of blacked-out rooms; and the infamous 'beep test' in which you are required to run back and forth between two points at an ever increasing speed. It was expected that any fear or physical weakness would be exposed during this process. Justin passed again, and was then faced with a series of interviews conducted by panels of firefighters, recruiting company officials, union members, and an ACT Brigade Superintendent. During this process, an obvious topic soon arose when an interviewer noticed Justin's wedding ring.

'So, you're married?'

'Yes,' replied Justin.

'Any children?'

'We had one but we lost her.'

'Tell us a bit about that.'

It was time for Justin to confront the most painful episode of his life. To be a firefighter one had to accept that death and trauma were part of the job. Justin needed to convince everyone, not least himself, that he was prepared for what lay ahead.

'I told them I'd been through fifteen months of hell on earth, and thought I could handle most things. Your experiences make you who you are and prepare you for what's to come. That's why I thought I would make a good firefighter. The job reflected so much about my life. Any traumatic incident is going to affect you, but how you handle it is the telling factor. Growing up on the farm, then suffering through Georgia, I had been around death and the pain it causes for a long time. I was aware of it. I had learnt that death is very much a part of life.'

It was with some disbelief that Justin read the letter he received after the interviews. This part-time farm contractor, shearer, cellar worker, baggage handler, movement control operator with an arts degree was about to launch a new career at Recruit College Number 25 of the ACT Fire Brigade.

Justin will never forget his first day of College. It was Thursday 7 February 2002. The true symbolism of this new beginning didn't become apparent until three o'clock in the afternoon when his mobile phone rang in the middle of a class. He was immediately taken outside by an instructor, but any admonishment from officialdom was lost in the words that Justin anxiously listened to on the phone. It was Sarah.

'They're getting closer. Only five minutes apart. I'm going to drive myself.'

'Okay, ring me back when you get to the hospital. I'll meet you there.'

After being assured by Sarah that he had time to finish his classes for the day, Justin anxiously watched the clock. At 4.15 pm he received another call. He was out the door in a matter of minutes and hurrying to John James Hospital. He arrived at 5 o'clock. Seventeen minutes later, Chloe Lilian was born, with her father holding her mother's hand, and still wearing 'the blues', the official uniform of the ACT Fire Brigade.

Although it was a time for indisputable joy, there was an undercurrent of apprehension. After the tragedy of Georgia's death, Sarah had been told by specialists that she had a one-in-twenty chance of having another child with a heart condition. It was only after a paediatrician thoroughly examined Chloe that Sarah and Justin could savour the joy of their daughter's birth.

But grief for the Morrisons was not far away. Tony's sister, Dell, died of cancer soon afterwards, prompting Tony, emotionally wounded again, to curse: 'I'm sick of burying Morrison women.'

It was during this time that Justin was studying at College 25. At 34 years of age he was one of the oldest of the sixteen recruits. He had started this new journey with trepidation. The list of things to learn was long: water pressure; pump and truck mechanics; the maintenance and use of breathing apparatus; driver training; road accident rescue; first aid; hazardous materials; decontamination; alarms and sprinklers; the character of fire . . .

After sixteen weeks he strode out of college carrying a fireman's glove and the number 25 attached to a handmade wooden shield. It was a reward from his peers; Fourth Class Firefighter Justin Morrison had been chosen as the 'Probies' Probie' (Probationaries' Probationary), similar to a 'players' player' on the sporting field.

As part of his initiation period Justin spent a month at each of Canberra's nine fire stations. Just three months into the job, he faced the moment that every emergency services worker inevitably confronts. It was a test of his character, a challenge to his emotions, which was magnified by his own history.

'It was a motor vehicle accident on a Saturday afternoon,' Justin recalls. 'We were first on the scene. There were two cars involved. I went to a bloke whose head had smashed against the steering wheel. Maybe I was lucky that I didn't go to the other car because it had a mother and daughter inside. I wasn't close enough to see them being worked on. Had I gone that way, it could have been totally different. All three ended up dying. A lot of blokes say that with fatalities, "Whatever you do, don't look them in the eyes." I went away thinking the accident could have been so easily avoided. I was definitely affected, but I got through. The main thing was that I wasn't going to take it home with me. It's something that I will never do because I want to protect my family from it. I have to draw the line and say "I can't do any more than I did."'

In the years since then, Justin has experienced other tragedies that have had the potential to push him into his past. One of his most vivid recollections is of a motor vehicle accident in which a pregnant woman was killed. Her baby was delivered by emergency caesarean but died shortly afterwards. Justin accepts such sadness is simply part of his job. In reality, it is part of his life. Yet in the four years that he has been a firefighter (at the time of writing he works at Ainslie, one of Canberra's busiest stations) he has also seen and felt the exhilarating flipside, those moments when people, possessions and property are saved. And then there is the absurd, such as the day that a young woman playfully flashed her breasts at a fire truck in which Justin was travelling. Firefighters see it all.

The difference for Justin is that in the last 'handful of years' he has seen more than most without even wearing his uniform.

***

On 26 November 2002, Tony Morrison woke up and put his work clothes on. Shearing week had just finished, the smell of lanoline was still fresh in his senses. He took a few steps, but felt 'a bit crook', so he returned to bed, not knowing that he would never feel another fleece, nor walk again on 'Rob Roy'. He died that morning of heart disease. He was 68. When police came to confirm the death, they sifted through the clothes he was still wearing. They found a tin of tobacco, another one of butts, a box of matches, a hanky, and a dry bit of sheep skin.

'We always said we'd cart him off that farm in a box, and that's exactly what we did. At least he didn't have to bury another Morrison woman,' says Justin smiling softly.

On 30 May 2004, Catherine Morrison turned 69. Three days later she went into hospital suffering from bowel cancer. She died on 17 September. Just nine days after her funeral, the cycle of Morrison life turned again when Justin and Sarah celebrated the arrival of their third child, a healthy boy, Joel Anthony. After all the love and support Catherine had given her family, it was a tragedy that she missed seeing and holding her new and only grandson by just over two weeks.

These days the children visit 'Rob Roy' regularly. Gerard has taken over the running of the property, but the whole family pitches in when needed. One day Justin wants to take Sarah and the children back to the land and continue a tradition that spans more than 100 years. Maybe they will break in another 'Carry Grog', or perhaps they'll discover a dusty tobacco pouch in the corner of the woolshed. However, for the moment they remain in Canberra's suburbs in a modest house with a living room that's warmed by family photos and a print of Ned Kelly, a man Justin admires because 'whether right or wrong, he was an underdog'.

The bearded face of Australia's most infamous bushranger hangs in one corner of the room. Yet it's not the portrait that commands most attention, but the words: 'Such is life.'

THE STEVENSONS
'Family Snapshots'

'Ambulance is family.'

Dave Stevenson, retired Paramedic, SA Ambulance Service

Ambulance Officer Dave Stevenson stopped at the crash scene. What had happened this time? No seatbelts? Foolish bravado? Weak brakes? Too many beers at the local? Whatever the reasons, the results were often the same. Death, injuries, tears. He'd attended motor-vehicle accidents at which bodies had been flung over fences, even up trees. He'd heard, seen, felt, and smelled death and distress. It was awful. But this situation was different from any other he'd been in. Instead of reaching into the back of his vehicle for his bag of equipment, he found a blanket to place over the cargo he often carried when off-duty. Then, he and his wife Kay walked calmly to the accident site, while behind them, the blanket moved just enough to allow three young sets of bright and curious eyes to sneak a look.

When that incident happened some 30 years ago, Dave Stevenson was briefly drawn away from his role as a father protecting his children: Kerrie, Chris and Pat, to one of an ambulance officer helping victims.

'I always remember the bodies hanging over the bonnet,' recalls Kerrie Shepperd. 'We saw a few accidents when we were growing up. Dad never came home and talked about work. What we knew about his job was what we saw.'

Whether it's through coincidence, bloodlines, social conditioning, fate or an intriguing mix of all four, Kerrie and Pat are still attending accidents today. The blankets have been shed, replaced by the green and silver uniforms that are worn by paramedics in the SA Ambulance Service. Like father like son, like father like daughter, like brother like sister. But the ambulance lineage doesn't stop there, as both Kerrie and Pat have married into the service. Like husband like wife, like sister-in-law, like brother-in-law. In all, five paramedics spanning two generations and four decades.

'At least they all have in-depth insights into nursing homes,' says Dave. 'So if I do have my stroke and survive it, I think they'll know where to put me.'

His laughter crackles through a throat that has loaned its bass chords to choirs and singing groups throughout Adelaide. He is a solidly built man in his sixties. He has strong hands, just a scattering of grey hair, and a gentle face that belies all it has seen.

Dave's journey to community service began on a suburban London street 40 years ago after a motorcyclist ploughed into a rubbish bin outside the Stevensons' home. Dave and Kay hurried out to see how they could help, but as they saw 'this poor bloke lying there with a broken leg', Dave realised how ill equipped he was for the situation. The English toolmaker thought little more of the incident until a few years later, when he was at a pub in Tea Tree Gully on the north-eastern fringe of Adelaide. At this stage he hadn't long migrated to Australia with his young family. All three children had been born by then, with ten-month-old Kerrie the youngest. While conditioning himself to the new taste and temperature of Aussie beer, he met a group of ambulance volunteers who'd just finished a meeting at the station next door, and they so aroused his interest and curiosity that within a few weeks he'd joined their ranks. All he needed was a senior first aid certificate, a clear driver's licence and no criminal record.

Working as a toolmaker by day, Dave became a volunteer at night, and soon learnt that there was no subtle introduction to a job that was primarily fed by human suffering. It was the early 1970s, the era of 'swoop and scoop' where officers arrived at a scene, quickly loaded the patient into the ambulance, then rushed him to the nearest medical aid.

Despite the traumas he saw, Dave found his enthusiasm for ambulance work was growing, and by 1974, he had downed the tools of his original trade, and had chosen a new career. His entry into the ambulance profession had happened by accident. A horrible accident. While still a volunteer he attended a double road fatality in which the victims were both children. Unbeknownst to him, he was being watched from afar by a Chief Supervisor. The next day he received a phone call from an operations manager at Ambulance Headquarters.

'Do you want to be a full-time officer?' asked the manager.

In the wake of answering 'yes', Dave entered a 'very blokey service where the coping mechanism was the pub'. In the following years he tried to filter the shocking events out of his mind, but the sheer weight of them ensured some would plague his memories, such as his recollection of the days that he carried away murdered bodies from crime scenes (these days this role is primarily performed by the Coroner's Department). Then there were the accidents involving children, none more dreadful than the one in which a newspaper boy standing on a street corner had been clipped by a truck and pulled under the wheels. He died. The sight of his bloodied body prompted Dave to think of his own children, most notably Chris who was about the same age as the newspaper boy. It immediately personalised the situation. A different place, a different situation, and Dave was aware that the child he saw could have been his own.

Dealing with the reactions of a victim's relatives became the most difficult part of his job. When attending an accident, particularly those of children, he always hoped to have the victim on the way to hospital before any distraught person arrived at the scene. He obviously felt for the people involved, but knew his job was made easier when there was no one crying on his shoulder. Despite those inevitable moments, Dave really enjoyed his job, believing he was in a 'privileged but perverse' position to intervene in people's lives. This commitment and passion took him beyond our shores to East Timor where sadness had bled into the culture. He recalls:

'There is one moment there that I live with every day of my life. We had this poor man who had walked for a day-and-a-half out of the bush to raise the alarm that his wife was having problems in childbirth. So my driver and I took him in the ambulance. It took us about eight hours to get to this village in the mountains. It was dark. I saw a man outside a hut hacking on a piece of wood. When I looked a bit closer I could see that he was making a small coffin. I went inside the hut. Not only was the baby dead, but the mother was as well. Everyone was looking at me expecting me to take control.

What do you do? I closed the baby's eyes, put the stethoscope on, went through the motions, baptised her, then carried on. It was very, very, very sad. You don't forget things like that.'

Dealing with death taught Dave how temporary life is. Although not a religious man, he was occasionally prompted to ponder about the afterlife. Certainly there were people and moments in his job that made him wonder, none more than the man who was adamant he'd left his own body while being resuscitated from cardiac arrest. He claimed he'd heard the words: 'Hit him again!', then felt the pain in his chest before going back inside his body. Mind games? Or fact? Dave can't offer an answer.

Reflecting on his 31 years in the service, Dave acknowledges he was part of a critical evolution. He witnessed simple but telling changes, such as the introduction of gloves to handle patients; he also saw the incredible developments highlighted by the life-saving clinical techniques now applied at accident scenes. He says the profession he first knew is now covered in dust.

'If an ambulance officer trained in the 70s suddenly arrived back on the scene today without any knowledge of the changes, he'd probably be horrified. There's no more swoop and scoop, no elaborate splinting of fractures. The emphasis is on keeping the vital organs alive, whereas before we didn't have the knowledge and technology to do that. In the 70s you'd go to the home of someone who'd had a cardiac arrest. You'd perform CPR, everyone thought you were wonderful, then you'd take the body to the mortuary. But now, with the monitoring equipment, the drugs, and the skill of the paramedics, patients have a much greater chance of survival. When you look at the face of a paramedic who has pulled someone out of a cardiac arrest, and they're confident that person will eventually walk out of hospital, the look on that face is priceless. You see it all the time. The public wouldn't recognise it, but I do.'

Although he never pushed any of his children towards the ambulance way of life, he admits he's 'quietly thrilled' that Kerrie and Pat followed his path. While Chris chose to travel the information technology superhighway (he currently works in the Indian computer capital, Bangalore), Kerrie never wanted to do anything else. From the time she was a teenager and began thinking seriously about her future, the desire to meet and help people loomed large in her thoughts. When she first told her father of her ambitions, the response was anything but positive.

'No! Don't do that!'

It was at a time when there were still very few women in the service, and Dave was concerned that at only seventeen, his daughter shouldn't limit herself to just one option. But Kerrie was determined. She did volunteer work for a year after school, then applied without telling her father. He was still unaware until he arrived back from an overseas trip and Kerrie greeted him with the news that she'd been accepted into the ambulance service.

Despite his surprise, Dave offered his full support, while Kerrie began an enriching career not only helping others, but learning from them. She says:

'You learn from other people's bad fortunes. Even the small things that most of us take for granted, like we might walk into a house in a lower socioeconomic area and see that the beds have no linen. You learn by looking and listening. I remember one job nearing Anzac Day. We attended to an old bloke who didn't want to go to hospital because he was waiting for his march, but he just had to go in. I rolled up his sleeve and there were numbers tattooed on his arm; he had been in a concentration camp. You could see the tears welling up in his eyes as he recalled details of it. You meet so many variables in this job. People are so different. You see how vulnerable people can be, and if you can put a smile on their face at that moment of need, then that's sometimes all the reward you need.'

As happened with her father, Kerrie discovered that parenthood, combined with her job, made clear what was important in life. She first met her future husband, Mark Shepperd, when they were both working on the road, but it wasn't until they met up during separate holidays in Europe that the relationship developed. They married in 2001. Three years later, baby Jack was born, and suddenly Kerrie had a much deeper understanding about what it means to care for others. There is one particular moment in her career that has ensured she will always consider raising a child as the most important duty she will ever have.

'We got sent to a house for an "unconscious collapse" and arrived to find a little girl and her father, who was drunk in the kitchen. He pointed to his wife in the lounge room. She'd been dead for probably twelve hours, but the little girl had no idea. She was the one who'd actually called the ambulance. She was there saying: "Wake up Mummy, wake up." We had to take her into a side room. She was only four or five, her pyjamas were saturated with urine, and her mattress had no bedding on it. It was saturated too. All her clothes were; she had nothing to put on. She was hungry, but there was nothing in the fridge except beer. That day I just wanted to pick that little girl up and take her home. It really rips your heart out. That incident made me want to try and make things better for every person I came across. Most of all, I've got from it an appreciation of my responsibilities to Jack. I hope he will never suffer anything like that little girl. She is often in the back of my mind.'

It's incomprehensible for the average person to understand what ambulance officers and paramedics see every day. A few years ago Dave was at a walkers' group meeting in Adelaide. He began talking with an army officer who was about to take some troops to Rwanda, one of too many African nations torn by internal conflict. Dave was surprised when the officer asked:

'Is it possible to bring some of the soldiers out on the road with you?'

'Why?'

'Because they're going overseas to work in a disaster area, but none of them have been exposed to death and trauma like you have.'

Ambulance officers and paramedics are indeed at the front line of our society, a front line that for Pat Stevenson is shaped by extreme dangers. Two years older than his sister, Pat was the first sibling into the service, but second onto the road. He'd begun as a storeperson, hoping that shifting boxes and running deliveries was just a transitory stage to joining the police force. However, a knee injury from football hindered his chances, and he was forced to reassess his options. Although he considered ambos to be 'blood and guts people', his views gradually changed as he met more officers. The athletic one of the Stevensons, he sought a position that would satisfy his love for action. He had been a world age junior trampoline champion, and with his lithe six-foot-four frame and hard-edged determination that was moulded by his sporting background, he offered the ambulance service someone outside the regular fit. So it wasn't surprising when he eventually leapt far beyond the normal roles and into a rescue paramedic position with the Special Operations Team.

Although every rescue had its unique difficulties, an autumn night six years ago threw down challenges that Pat had never encountered before, and, he hopes, will never face again. It began when an alert came through of a vessel in distress off the Yorke Peninsula west of Adelaide. Pat, with two pilots and an air crewman from Lloyds Helicopters was soon hovering over a bedraggled group of men and women clinging to ropes tied to the hull of the overturned racing yacht *Doctel Rager*. It was about 11 pm: the darkness, rising swells, a strengthening wind, and the daunting presence of sheer limestone cliffs all added urgency to the situation. But this wasn't all. Pat hadn't worked with this rescue crew before, and the helicopter had just undergone eight weeks of refurbishment before it was due for deployment in Perth. Test flights had only finished shortly before the call-out; it was a 'stroke of luck' that the rescue craft was still available.

In many rescues, it was standard procedure for a crew to survey the site and consider its options before beginning any operation, but on this occasion, there was no time for deep assessments. Lost minutes meant lives at greater risk. By this stage it had been about three hours since the initial EPIRB (Emergency Position Indicating Radiobeacon) distress signal had been received. During this period, the boat's hull had sunk further into the water, and as every wave crashed over it crew members were being pulled back on top of people after washing into the sea. They were cold, some were suffering extreme hypothermia, and the chances of a life being lost were increasing.

Despite all his experience and training, nothing could wholly prepare Pat for what he had to do. The boat and people looked so tiny, the task so enormous. He rigged up his harness, checked all the fittings, nodded to his crew, and was then lowered into the night at the end of a safety winch wire. Twenty metres below, and the same dangerous distance from the rocks and cliff face, twelve desperate people (seven men and five women) watched, partially blinded by the helicopter's lights.

Less than a year earlier, Pat had led a helicopter rescue training course at South Australia's Cruising Yacht Club. The same people he'd saved in mock rescues were now students of reality; if ever there was a time to adhere to all they had learnt, this was the moment.

Pat swam a few metres to the boat, and motioned to the boat crew for one member to come with him. Within seconds a woman had dropped into the water, and was being supported by Pat as he attached the safety strop. He raised his hands and gave the thumbs up to his crewmen. The line tightened, then pulled its load upwards. The ordeal would soon be over for the woman, but not for her rescuer. One down, eleven to go.

Down and up. Down and up. Down and up. Down and up. Four more people were plucked from the sea before the chopper needed to return to land for re-fuelling. At least Pat, already exhausted, could rest briefly. But for the seven people remaining on the boat, the situation was worsening. The 17-metre yacht was breaking apart under the constant pounding of the waves. With the wind lifting to 30 knots, there seemed every chance that it could smash into the rocks.

When the chopper returned twenty minutes later, the silhouette of a familiar figure was cut against the helicopter's lights. Each time Pat hit the water, he braced his body for another bashing and dunking. This was perhaps the hardest part of the job, as he became disorientated when he went below the water, and needed to gain his bearings each time he surfaced. The salt spray swirled and stung, and the waves thundered, making gasps of air difficult to take, but mouthfuls of water painfully easy.

Down and up. Down and up. Down and up...

It was all over in another hour. Pat sat back in the chopper, emotionally and physically spent. Twelve times he'd dropped into the darkness and slapped into the sea; twelve times he'd churned through the swell to reach the boat; twelve times he hitched a desperate person to the safety strop; and twelve times he supported his relieved load as the wire shortened and lines of relief lengthened. He recalls:

'The rescue was full of raw emotion and adrenaline. I still get a lump in the throat thinking about it. I was knackered after lifting the first five, and didn't know how I'd go with the other seven, but they proved easier than I thought. In something like that it's easy to have self-doubts, but you have no choice; you have to keep going. It just goes to show what the mind and body can do. You could see the thanks in their eyes. It's one of those times you look back on and feel proud of yourself, your workmates and what you represent.'

Most emergency service personnel don't seek recognition for what they do, but nevertheless, acknowledgement can be heartwarming. In the weeks after the incident Pat received letters and cards from many of the people he had rescued. He was made an honorary member of the Cruising Yacht Club of South Australia, and was re-introduced to the people he'd saved. In such moments, there is no more powerful word in the language than 'thank you'. Later he was awarded the Ambulance Service Medal.

Considering the course already mapped by his father and sister, Pat surprised no one when he married Sonya, an Intensive Care Paramedic. It was an ambulance cliché—they met on the job, although there were some obstacles to overcome because in the early years of service Pat firmly vowed: 'I'm never gonna date an ambo!' They now have two children, Aidan and Brodie.

'It's not at all surprising there are marriages in this job,' acknowledges Dave. 'You share the same traumas, go to jobs together, share the same rooms when you're waiting for jobs. It's the ultimate sense of being a team. I've been to those team-building courses but they wouldn't come close to the rapport that happens between an ambulance crew. They stand together through thick and thin.'

Dave retired from the Ambulance Service in 2005. His tireless efforts hadn't come without a cost. He is now divorced, but remains married to community service, and since signing off from official duties, he has travelled to Sri Lanka to help improve ambulance facilities there. Perhaps age will eventually apply the brakes to his drive, but if that happens, there will be no shortage of other Stevensons to take over. And rest assured, family reunions will never run dry of stories.

DEB WALLACE

'The People Person'

'We all have to go over speed bumps in our life, but when you fall, just pick yourself up, dust yourself off and keep going.'

Detective Superintendent Deb Wallace, New South Wales Police Service

Deb Wallace walked smiling out of the Parramatta High exam hall. What a relief! The Higher School Certificate was over, and it was time to forget all the theories and definitions that she'd crammed into her mind for the sake of marks and percentile bands. The only subject that mattered now was maths, and counting enough coins to pay for bus and ferry rides to Manly Beach. As for the future? Well, maybe she'd become a travel agent, or a dentist's assistant, but she was in no hurry; the pursuit of a job wasn't as enticing as chasing the summer sun.

Her father thought otherwise. Ken Wallace was a gruff but loving man of few words. He was a security locksmith at

Prospect County Council in Sydney's western suburbs, and he also had part-time jobs as a cleaner and taxidriver. His years of experience and the intricacies of his professions had taught him just how important it was to open doors. There was no time to waste. The day that Deb's exams finished was the day to pounce. It was a Wednesday.

'So what are you going to do?' he asked the third of his four children.

'I don't know. I'll think about it for a while,' replied Deb.

'Well, don't think too long because I've pulled a few strings and got you a job at the County Council. You start Monday.'

'Doing what?'

'Clerk typist.'

Ken paused, allowing his wonderful news to sink in. Then, he took a breath and continued:

'It's a good job, and in fact if you play your cards right you could be there forever because it's a government job and it's secure!'

Deb looked at him, trying to hide her surprise. She adored her Dad, and wouldn't dare tell him what she was thinking: 'Oh my God! I'm eighteen and this is what's ahead of me!'

Five days later, Deb somewhat reluctantly began her career as a VDU (Visual Display Unit) Operator who spent much of her time answering phone calls with the standard welcome: 'Prospect County Council, may I help you?'

But she was forever asking herself another question: 'What do I want to do with my life?'

She was encouraged by her mother, Aileen, who wanted her children to look far beyond the bricks and blocks of 1980s suburbia. She was forever warning Deb: 'Don't follow what I've done. I've lived inside four walls while raising four children, and your father has three jobs. Don't make the same mistake. Find a life for you.'

Deb's work colleagues, most of whom were married middle-aged women, offered similar advice. They recognised that their young associate wasn't the type of person to want a wedding ring on the finger, and a house that was squeezed into the mortgage belt.

Deb wanted to explore. She wanted adventure. But how? After she'd spent nearly four years sitting behind a desk and a terminal, a possible answer was pushed under her nose. Her colleagues had noticed two job advertisements in a newspaper, and promptly decided they'd appeal to Deb.

'Take your pick,' they told her. 'You can't stay here. This is your chance to get away and do something.'

Both Qantas and the New South Wales Police Service were seeking new personnel. Deb was so intrigued by the ads that she seriously, and quickly, considered her options. She could stay where she was and risk becoming Miss Monotony? No, it was definitely time to move on. She could apply to become an air stewardess, serving hundreds of people at thirty thousand feet above sea level and travelling the world? No, despite being a slender, leggy brunette with a bubbly nature, she didn't consider herself attractive and well-groomed enough to 'look like a model'. So, she was left with the Police Service, where she could serve whole communities with her feet firmly on the ground. Yes, she would apply. Her decision was somewhat influenced by the knowledge that her older sister, Anne, had a friend who was a police officer, and she was 'the most elegant woman' that Deb could remember ever having seen.

When she told her family of her plans, Deb watched her father's eyes roll back and his mouth tighten in the silent but readable expression: *'Oh no, we're in trouble now!'*

***

Little more than a month later, 22-year-old Deb stood at the foot of a steep grassy slope in Sydney's Moore Park. She was one of just a handful of women in the 100-strong 'Class 194' of the New South Wales Police Academy, which was then based just minutes away in the inner-city suburb, Redfern. The slope, about 80 metres from top to bottom, was known by recruits as 'Breakfast Hill', as it was a common place to heave and leave your cornflakes, eggs and toast. Deb linked arms with the recruits on either side of her in the middle of a long chain of determination. It surged upwards, each step harder than the last until the summit was reached. This was all watched on by the steely eyed fitness instructor, Brian 'Chikka' Moore, an ex first grade rugby league player, whose bald head glinted in the morning sun. Days earlier he'd shown his rookies footage of a woman competing at the torturous Hawaii Ironman, a day-long test of endurance and willpower. Approaching the finish, the woman wobbled, fell to her knees, crawled across the line and then collapsed. When 'Chikka' turned the lights back on in the auditorium, Deb could have sworn she saw a tear in her instructor's eye as he told his troops: 'I've seen that hundreds of times before but never failed to be inspired by it. That's the level of commitment I want from you.'

Afterwards Deb marched out onto the parade ground thinking: 'What have I done?'

Her resolve was soon tested when the recruits were sent on a 10-kilometre run that made the rigours of Breakfast Hill seem like a palatable entrée. As she shuffled along, Deb was startled when Chikka's voice boomed over her shoulder:

'Trainee Wallace, you're not going to make it if you keep going like that!'

Choosing silliness over discretion Deb replied: 'Sergeant, I've got it all worked out. When I really start to struggle I'll lose control of all my bodily functions and slump to the ground. Then you'll start to cry in public and carry me back.'

Chikka wasn't impressed.

Despite occasionally raising the ire of her instructors, Deb graduated from the Academy, and was posted to Blacktown in Sydney's western suburbs. Despite entering a male-dominated service, she was immediately welcomed at an eternally hectic station where: 'You were a number and you counted.' It certainly helped that the station's charge-room at the time was 'run like a military camp' by Second Class Sergeant Joan Stedman, who impressed Deb because she didn't compromise her femininity and integrity. She became one of PC Wallace's many role models.

When Deb first arrived at Blacktown, she was frequently rostered on to patrol car 27–3. 27 referred to Blacktown's station number and 3 was the vehicle's number. 27–3, which normally had a sergeant and one of the junior constables turned up to incidents after 27–1 and 27–2, which were the front line Ford F100 trucks that were appointed to the senior constables. As time passed Deb was thrilled to see her name on the roster whiteboard appearing more regularly in the 1 and 2 columns.

When at the front line Deb faced the additional challenge of being a female in sometimes violent situations dominated by males. This was never more obvious than when dealing with fights that were fuelled by alcohol. By the time police arrived, the altercations were often over, and there was rarely ever a need to discover the answer to the testosterone-filled conundrum: 'What use is a bird in a brawl?' However, on other occasions Deb turned up to find that she could help defuse a situation simply because of her sex. Amid the bloodied noses, flying fists and great Australian adjectives, it seemed some fighters were still gentlemen.

'Sorry, love! Fellas, there's a woman here. Watch your language!'

Perhaps one of Deb's biggest gender worries came from home. Her father was in a fighting mood after he discovered someone was stealing the Wallace milk money that was left outside the front door each night. Ken resorted to sleeping with a saucepan beside his bed in the hope that he'd hear the thief and be able to spring to his feet and clobber the culprit. Deb tried pleading with him.

'Dad, call me. I'm a police officer now.'

But Ken just rolled his eyes as though to say 'you're kidding aren't you!' The milk money bandit was eventually scared off after he was almost skewered by a curtain rod that Aileen hurled at him. It seemed Wallace women could look after themselves.

However, Deb was still grateful that she had a good bloke to guide her through the maze of her early days at Blacktown, a station which when it wasn't busy, was frantic. As with every probationary constable she was appointed a 'buddie'. Senior Constable Ron (Ronnie) McGown was a short, stocky man whose sense of duty for his job was matched by his commitment to his new colleague.

Appearing in court was among Deb's greatest fears. She'd never done public speaking before, and dreaded the day when she'd finally have to give evidence in a 'Not Guilty' case. Ronnie eased her towards that occasion by involving her in simple matters, such as shoplifting cases, where the offender pleaded guilty. After charging the culprit, it was generally a matter of taking him or her to court the same day. It was here that, after seeking the magistrate's permission, Ronnie allowed Deb into the witness box to read the facts. This was at a time when police relied on typed records derived from notes taken during interviews; the electronic age of interviewing was still a few years away. This note-taking inevitably led to accusations of 'verbals', where offenders who pleaded 'Not Guilty' accused the police of either misconstruing or making up what was said. Consequently, defence counsels regularly tried to discredit police evidence. It could be a daunting affair standing in a witness box having your professional reputation picked and occasionally torn apart. Because of Ronnie, Deb had completely overcome her apprehension when that day finally did arrive.

However, nothing could wholly prepare her for the night a clichéd movie script leapt into reality. She was on general duties with another officer when the call crackled through on their truck radio. There had been an armed robbery at a service station. The two assailants, reported as having a sawn-off shotgun, had fled from the scene in a car. Deb and her partner went to the service station to obtain details from the employee who'd been held up. His jittery recollections included the revelation that one of the robbers had yelled to his partner a suggested road on which to dump their car. Blacktown police car 27–3 then headed off to that location with its officers wary but not convinced they'd find anything unusual. Surely the robbers couldn't have been so stupid as to have announced their real plans. When Deb and her partner turned onto the mentioned road their attitude changed abruptly. A corridor of factories and fences led to a car fitting the description given by the service station worker. It was parked at a dead-end near a creek. 27–3 edged closer, and when its headlights split the darkness and shone directly at the car, Deb shifted in her seat as she thought: 'Oh God, I hope they're not there.'

But they were. Two men stood in front of their car just 25 metres away. After her partner relayed the information over the radio, Deb stepped out and pulled her gun.

'Stop! Police!'

It was more of a squeak than a demand. Everything seemed to happen in slow motion. As Deb tried to stop her body from shaking, she watched one of the men move an arm as though he was reaching for something. *Oh God!* Deb glanced at her partner, who was standing on the other side of the patrol car. Neither knew if they were a trigger touch away from falling, as they couldn't see a gun. In the murky distance, the two men looked at each other. Four people crowded in a frozen moment in an otherwise deserted back-street.

The men turned and bolted along the creek. Deb still couldn't see if they were armed. She went back to the wagon and reached for the radio, her hands trembling, and uncertainty swirling in her thoughts. Should she announce over the airwaves that the assailants had a weapon? If she did, what would happen if one of her colleagues fired a shot on the strength of her information only to find that the offenders had nothing? What? If? Why? How? When? Should she? Shouldn't she? Questions and possibilities sprinted through her mind in a matter of seconds. She buttoned on to the radio and squeaked again. She gave her location, and the direction the suspects were heading on foot. It was all standard information, but as for the possibility of a weapon, she could only announce:

'For all cars attending, one of the assailants has something long and hard in the front of his pants.'

Silence followed, leaving Constable Wallace to wonder what she had said. Then she heard the familiar voice over the airwaves of her buddie Ronnie:

'27–1 radio. Could 27–3 confirm if the long hard object was cocked?'

The robbers were eventually captured. Now, more than twenty years later, Detective Superintendent Deb Wallace can't recall whether or not they had a gun. Too many cases have flashed by over the years to recall every detail, but she says it's impossible to forget how she felt.

'I must have looked like a human earthquake, shaking from the tip of my head down to my shoes. For those few seconds, apprehension and fear for the future went through my mind. I honestly didn't know if I had a future. But afterwards it was wonderful because I copped a lot of banter and jokes. I love the police spirit.'

But that spirit is often challenged by the hardships that officers are forced to regularly endure. Each officer has his or her own way of coping with fatalities and trauma, and the subsequent reactions of a victim's family and friends. Deb was introduced to death when attending motor vehicle accidents. She was often taken by the silence that accompanied a fatality, especially in Sydney's western rural fringes, where she arrived to find the stillness was only broken by voices scratching on the police radio.

In February 1986, the quietness of the bush was battered by a death that not only changed Deb's career direction, but affected her life. Constable Wallace, nearing three years as a general duties officer, was working the switchboard at Blacktown Station on the evening that Garry Lynch walked in to report his daughter missing. She hadn't returned home after going out to dinner with friends in the city the previous night. Nor had she turned up to Sydney Hospital for a nursing shift just hours before her father approached the police. The next day, her naked body was found in a cow paddock in Prospect, a few kilometres away from Blacktown. Her throat had been sliced, her face was severely swollen, and she had a number of fractures. Further examinations revealed that she'd been raped. The full list of her injuries was grotesque.

Five days later Deb walked into the spotlight of one of the most publicised murder investigations in Australian history, the Anita Cobby case. With a specially gathered Task Force still to find the killer, or killers, it was decided to re-enact Anita's movements the night that she was last seen alive. Deb was chosen to play the role. She was the same age as Anita, and of similar appearance.

On Sunday 9 February Deb arrived in a police car at Eddy Avenue, outside Sydney's Central Station. A week earlier, Anita had been dropped off at this point by one of her friends after dinner. Deb was dressed in similar clothes to the ones Anita had been wearing: ski pants and a singlet. Deb also had a black belt, and a large over-the-shoulder handbag. When she opened the car door, she was met by a media scrum.

Together with some detectives from the Task Force, and amid the cacophony of clicking shutters, flash explosions, and the shuffling shoes of TV cameramen walking backwards, Deb headed through Central to Platform 8 where she boarded the 9.12 pm train to Blacktown. Once the doors were shut, the detectives were away from the media and free to concentrate on interviewing each passenger in the hope of jogging the memory of anyone who may have caught the same train the previous Sunday.

Deb and her colleagues alighted at the other end to find more media waiting. After answering some questions she began walking the same streets that Anita had taken towards her parents' home. Even with police watching her every step, Deb recalls this was an eerie and intimidating experience.

'It was about a 30-minute walk. The purpose was to match the timings of witnesses who'd seen Anita walking the same route. Before I started, one of the detectives said to me: "I wonder if the killer is like a firebug, and has come back to watch you." That made me think. When I started, the detectives trailed about 200 metres behind me. It was dark and hot, and again it was the silence that I noticed. It was very quiet, and I began thinking that Anita wouldn't have known that she was perhaps into the last hours of her life. What was she feeling? The terror she must have felt at some point was very sobering for me.'

The next morning, Detective Sergeant Graeme Rosetta, joint head of the Task Force, asked Deb to join the investigation. She was primarily involved in discovering accessories to the crime, such as people who may have been told about it, or had destroyed evidence. All around her, determination and passion drove the team around the clock. Leads were dissected and acted on; no detail was overlooked. This was fastidious, painstaking, frustrating work. On several occasions Deb noticed Detective Sergeant Rosetta tapping into the early hours on his 'clang bang' typewriter.

This commitment was rewarded when five local men, Michael Murdoch, John Travers, and brothers Leslie, Gary and Michael Murphy, were arrested. As more details of their abduction, gang rape and murder of Anita were revealed, public rage spread all across the nation. There is no need in these pages to describe what happened to Anita other than to say there have been few more ghastly crimes committed anywhere. Sixteen months later, in June 1987, the five men were all given life sentences with the recommendation that they never be released. Deb was in court the day the guilty verdicts were handed down. As cheers broke out, she looked across at Detective Sergeant Rosetta. What would he do? Punch the air? Yell in triumph? He nodded, picked up his papers and quietly walked out of the courtroom.

Throughout the entire case, Deb never met Anita's parents. During the re-enactment she had felt 'embarrassed and intrusive about pretending to be their daughter'. She still held on to those sentiments at the trial, where she saw Garry and Grace 'displaying their strength every day in the courtroom'. Deb couldn't approach them, fearing she'd stir emotions. She thought it best to move on; the Lynches had already endured too much.

It took fourteen years for their paths to cross again. Deb had been appointed to a curatorium that was responsible for the quality control of 'Anita and Beyond', a sensitive exhibition that examined the Cobby case and its effect on the Western Sydney community. When she attended the first meeting at the Casula Powerhouse Arts and Crafts Centre, nerves gripped her after she noticed Garry and Grace were there. She needn't have worried. When an introduction came through a third party, Garry, by now into his eighties, stood and welcomed Deb as though they were old friends.

'G'day Deb. How are you going?'

With the warmth of a handshake and a smile, a friendship was formed that flourishes to this day. The three meet regularly, and occasionally talk about their journeys since Anita's death. Garry and Grace have never asked Deb why she didn't approach them all those years before. Answers to that aren't important. The past has taught them all that they must care for the present and hope for the future.

Until she'd become involved in the Cobby case, Deb couldn't imagine doing anything but working in general duties. She enjoyed the challenge, thrived on the adrenaline, and cherished the friendships. However, her exposure to detective work changed her views as she gained enormous admiration for the many people working behind the scenes. In 1986, three years after joining the police, Deb left the front line and moved into a detective's role as a plain-clothes constable at Blacktown. She was first partnered with Bobby Broad, a battle-hardened man who, with tongue firmly in cheek, accepted his new associate.

'I may as well have you because I have three daughters and a wife, so one more woman won't make a difference.'

But she did. In one case, in which they were accused of 'verbals', Deb and Bobby were grilled in court by a defence counsel who suggested the two detectives were 'closer than husband and wife'. In some ways they were; tight-knit detective units readily lent themselves to the marriage of 'work spouses'. In a partnership spanning nearly three years, Bobby set the standard for Deb. Sadly, he died many years later in a cycling accident. Deb, a pallbearer at his funeral, told his widow:

'It was an honour to carry Bobby because he carried me for a long time.'

By 1989, Deb was a Detective First Class Constable who was on the move. She said goodbye to Blacktown headed to nearby Cabramatta, one of Sydney's most colourful, vibrant, interesting, yet often troubled suburbs. In the late 1980s and 90s it was considered one of the most dangerous and volatile crime areas in Australia. It had a large Vietnamese community whose reputation was darkened by heroin dealing, extortion, and murders. The notorious 5T gang was at the centre of this infamy. When Deb started work here, she was abruptly introduced to a whole new way of policing.

'I was used to arrests where the process was normally: you'd have a chase, then the proverbial wrestle, you'd put the offender in a headlock; your partner would put the handcuffs on, then it was off to the back of the truck. Sometimes the offender might turn and shout out to his mates: "Tell Mum I'm at the cop shop."

'But I get to Cabramatta and it's very, very different. I go for a walk with Scott Cooke, a young constable who was a detective in training. I'm a little apprehensive about this place because of its tough reputation, with all the sinister people apparently lurking around. This Vietnamese man with long hair wanders over to us, and starts a conversation.

'"Hello, Mr Cooke."

'"Hello Duc, how are you today?"

'"Very good, sir. Who is this with you?"

"This is our new detective, Detective Wallace."

'Duc couldn't pronounce my name so he decided to call me "Madam", which they tell me is a very polite term. Cookey then gets serious.

'"Duc, mate, we have to talk to you about the extortion."

'"Which one?"

'"The one at such and such a restaurant."

'"No, I wasn't at that one."

'"Well mate, you've been identified, so you have to come down."

'At this time I'm thinking we're about to start doing the proverbial wrestle, but Cookey does it another way:

'"So, Duc, can you come down to the station this afternoon? I'm just taking Detective Wallace for a walk now."

'"What time, Mr Cooke?"

'"How about two?"

'"Okay."

'Then we go our separate ways and I said to Cookey: "Extortion is pretty serious! You don't really expect him to turn up, do you?" Cookey replied: "This is Cabramatta. This is how we do it." We go back and have lunch, and at five to two Duc rolls up, saying: "Hello, I'm a bit early, is that okay?"

'That set the tone for policing in Cabramatta. I learnt a lot from those very experienced and smart young constables, and I learnt a lot from the community too. Very early on I was given advice by a 5T member.

'"Madam, you can't beat us unless you understand us."'

Such understanding couldn't be learnt by burying a nose in books. For the following five years Deb spent more time on the street than she ever had before, in what were some of the 'toughest, most interesting, challenging and rewarding years' of her career. Surprisingly many gang members in the area had a certain respect for the police, acknowledging the age-old adage: 'If we do the crime, we will do the time.' However, for every offender sent to prison, there always seemed to be many more to take their place.

In the early to mid 1990s Cabramatta changed character as more and more outsiders were lured in by heroin. Caught in the cycle of needing money for their fixes, they lifted the crime rate. This exposed the police to some curious behaviour by 5T members, who were responsible for some horrendous crimes, yet would step onto the other side of the law when they saw some petty offences committed. There were times when in high-heeled pursuit of a bag-snatcher, Deb was passed by sprinting 5T members yelling: 'Don't worry, Madam, we'll catch them for you.'

This was an environment for both the absurd and the savage. Cabramatta hosted some of the most infamous crimes in Sydney's recent history including the assassination of local State parliamentarian John Newman. 5T leader Tri Minh Tran was also murdered, prompting one of his followers to tell Deb: 'We are a snake without a head. Our brain has been cut off and we have no direction.'

In the aftermath of Tri's death, Deb was placed in charge of the Cabramatta Gang Squad, which was primarily responsible for dismantling the 5T's activities. Many operations were covert. In a bid to stop some vulnerable youngsters in the area from becoming the next generation of prison numbers, Deb approached Father Chris Riley whose 'Youth Off the Streets' projects have helped change the lives of thousands of troubled youths. Although not every initiative tried in Cabramatta was successful, some boys and young men who were heading in the wrong direction took U-turns and have remained out of trouble with the police to this day.

Deb's devotion to her job revealed the faint line that sometimes exists between right and wrong. She was accused of having an inappropriate association with some gang elements, and was consequently investigated by the New South Wales Royal Commission into Police Corruption. The allegations were quickly dismissed.

By 1998, she saw the flipside when she was appointed as an Acting Inspector in Internal Affairs. Investigating her colleagues brought 'some of the most disappointing and unpleasant moments' of her career, but as is her way, she prefers to look at the positives during these three years, and is proud that in addition to a 'very, very small minority of wrong-doers are many, many hard-working, dedicated people who put their hearts and souls into their jobs'.

After leaving Internal Affairs, Deb passed briefly through a number of positions before she answered the irresistible call from Cabramatta where she became Crime Manager. Although the general public's perception of police work often ends at guns, car-chases, and the ubiquitous CSI television programs, Deb became heavily involved in an area that may have lacked glamour but still offered a tremendous sense of satisfaction: policy development and legislation.

Because of the high presence of uniformed officers on Cabramatta's streets, drug dealers were forced to change their strategies. They moved away from the central business area into residential blocks, where they established fortified drug houses. These were distinguished by massive double reinforced steel doors. The dealers would make their exchanges with buyers through a small mesh hole in these doors. The dealers' faces were never seen, only their hands. Police surveillance operations readily identified these places, but by the time search warrants were obtained, and twenty minutes of sparks flew from chainsaws and other cutting equipment, the people behind the doors had got rid of the drugs and were sitting innocently playing cards.

Police conducted more than 50 searches over a few months. More than 50 doors were destroyed. But not one arrest was made.

Under the direction of Regional Commander Clive Small, Deb was part of a team that proposed new strategies which eventually led to a change in legislation. In 2001, the New South Wales State Parliament introduced a law that gave police the powers to charge culprits with a variety of offences, including being found on a drug premises, entering or leaving a drug premises, or organising a drug premises. They usually weren't able to issue the more serious charges concerning supply or possession, but the new law carried a hefty deterrent: first-time offenders could receive as long as a year's prison sentence. As a result, it took just a matter of months for most drug premises in Cabramatta to close down.

By this stage, Deb had moved comfortably into senior management. Her journey into the corridors of decision-makers took her to State Crime Command where for nine months she was acting Commander of the Firearms Squad. She was involved in developing policies and think tanks on how to get guns off the street.

Nowadays, as Detective Superintendent Wallace, she is Commander of the 80-strong South-East Asian Crime Unit, which focuses on organised Asian crime across New South Wales. It is based at Parramatta, just a few minutes drive from where Deb walked out of high school.

She certainly didn't follow the path that her father mapped out for her, but this has never affected their relationship. Ken is immensely proud of his daughter. That's not to say he doesn't think she still needs a little support every now again. His phone conversations with Deb have been known to end with: 'Bring your car over. I'll give it a wash for you.'

Deb remains very close to her parents, who've taught her 'It's not what people say that matters, it's what those closest to you do and think'. And for those who know Deb, her actions can never be questioned. If you are her friend she is the type of person you could imagine walking unannounced down your driveway, with a freshly made pavlova, a bottle of wine and a smile. She is a giver. When not working, she devotes her time to numerous charities, and periodically takes groups of women on shopping trips to Cabramatta. She is a happy person who laughs a lot, often at herself. She is also an eternal optimist who adheres to the saying of a colleague: 'Don't pole vault over mouse shit!' In other words, don't make issues of small matters.

Because her life is in perpetual motion Deb has never stopped long enough to get married. Nevertheless she has met 'many wonderful people' and has 'many wonderful friends'. She has lived and worked her entire life in Sydney's western suburbs, an area that is periodically mocked by others in wealthier suburbs. Deb doesn't care about such criticism. She is a 'westie through and through and proud of it'. However, the great beauty of her character is that it reaches out and touches people in all directions. To steal a likely line from the regimental Chikka Moore: 'Keep going, Deb. You haven't finished yet.'

GRAEME JONES

'Clinical Focus'

'Our uniform means we sometimes go where angels fear to tread.'

Graeme Jones, Paramedic, Tasmanian Ambulance Service

Graeme Robert Jones once worked for an insurance company. If he was to return to that role today, what premium would he set for a paramedic who quite readily risked his life for others?

When recounting his experiences, Graeme commands attention. He has a deep clear voice, and a strong presence defined by broad shoulders that have thrown many a racing motorbike around tight corners. Nearly all of his 50 years have been spent in and around Launceston, in north-east Tasmania. And since 1972 the years have been powered by a passion that perhaps is best revealed by what is yet to come. He says: 'It really does disturb me to think that in another ten years my time in the ambulance will be over. I wonder what

I could do that would ever replace this and give me the same fulfilment. I've never sat down and analysed it fully, but prior to marrying and having children, ambulance meant everything to me. It has always made me feel as though I'm worth something personally; it makes me feel better about myself.'

During his career he has been in situations that have demanded incredible courage, yet he has never allowed himself to lose focus. However, sometimes that focus creates additional dangers. On one occasion Graeme and his partner attended to a gunshot victim in a family home. There was no time to lose. They set to work: pulse, ventilation, blood loss, lines in . . . Focus. It was only minutes later they were distracted by the commotion of several people trying to convince a man to leave. It was the gunman. He'd been in the room holding the weapon ever since the paramedics had arrived. Graeme hadn't noticed him.

Graeme Jones has received several awards, including the Royal Humane Society's Bronze Medal for Bravery, and the Bravery Medal, which is part of Australia's National Honours List. He is very modest about such recognition.

'It's a strange feeling when you get commended for a job you've done. If it hadn't been me it would have been another officer doing it. It's great recognition for the ambulance service, not just me personally. I like to think that it's my profession that's been recognised.'

The following stories are two brief accounts of the work ambulance officers do at the front line. They are written in memory of Graeme's former partner, the late Roly Holton.

21 OCTOBER 1995

Paramedic Graeme Jones wound down the window and asked the police officer at the road block, 'Is it safe to go in?'

'I'm not sure.'

Graeme looked at his partner, Roly Holton, then stared through the windscreen. They had no choice. None at all. They had to try. About 300 metres ahead of them they looked at the policemen crouching down behind a fence. Another one lay motionless on the ground. Gunshots rang out.

A road was directly in front of the policemen. A block of units was beyond that. And somewhere inside was someone with at least one weapon.

Graeme thought: 'This could be ugly!'

What to do? An officer was down, and no one knew his condition. Graeme and Roly had to reach him. Somehow. They grabbed two bags each of emergency equipment, then sprinted across the road on which they'd parked. They jumped a fence that stretched towards a block of units that was diagonally behind the policemen taking cover. They ran towards the units, unaware of the gunman's position. In the broad daylight, they were easy targets.

They reached the units, sucking at air, hearts pounding. They dropped their bags. The policemen signalled them to stop and stay where they were, but this wasn't a time to obey the law. They looked at the wounded officer about 30 metres away, then glanced at each other.

'You ready?' asked Graeme

Roly nodded.

'Let's go.'

They ran, crouching forward, adrenaline powering every stride. Two policemen hurried towards them, pulling their bloodied colleague with them. The exchange was quick. Graeme and Roly took the officer by the shirt. He was conscious, his face contorted, he yelled in pain.

'Sorry, said Graeme. 'This'll hurt.'

They dragged him towards the units. The officer swore. They turned the corner and reached relative safety. There was a snippet of silence, then more shots.

The officer had been shot in the thigh and there was a risk he could haemorrhage to death because one of the body's biggest arteries, the femoral artery, was in this area. The paramedics did a quick assessment. Thankfully, no life-saving intervention was needed. But they couldn't stay where they were. More shots were fired. It was impossible to know where the bullets had come from, and crucially, where they were going. A decision was made for Roly to dash back and get the ambulance, then try to drive it to the units, keeping as close to the fence-line as possible. He hurried off while Graeme worked on the officer.

Roly returned and parked the ambulance behind the units. The paramedics prepared the officer for loading. They crouched, they crawled, they focused. And as they did, an elderly man stood watching them.

'Can I help you?' he asked.

'What! Where on earth did you come from?' replied Graeme.

'From the units.'

'Well, go back inside and lie on the floor. And don't move!'

The man nodded and headed back to his home, his cameo appearance adding a touch of absurdity to a volatile situation. Graeme and Roly returned to their task of loading their patient. Stretcher out, officer on, push, go hard, shut the doors, get inside, now *go*! Roly reversed the ambulance, and suddenly exposed its tail out from behind the units.

Bang!

A spray of shotgun pellets hit the back of the vehicle. The injured police officer dug himself into the stretcher mattress, Graeme lay on the floor expecting a more powerful shot to come bursting through at any moment, and all Roly could do was drive.

The ambulance was soon on its way to hospital, and the injured police officer was on the road to a full recovery. Graeme and Roly were acknowledged as heroes by the media and Launceston community, but Graeme simply says they were doing what they were trained to do:

'Our safety didn't really come into it. The patient has always been the centre of my attention; if I know someone is crook I will try to get to them. Yes, I would be summing up the dangers to myself, but I would still be trying to get to the patient. It's strange though. On that day I never felt scared. I just felt exhilarated.'

28 MAY 2002

The freezing air slapped the faces of the men in the seven-man rescue crew. There were police officers, SES volunteers, and Paramedic Graeme Jones. Branches crunched under their feet. It seemed they'd been walking for an eternity, each lugging 30-kilogram packs full of emergency medical aid. But would it be enough? They didn't know how many were injured, or what condition any victims would be in. All they could do was keep going, keep hoping. Step after step in the darkness through some of the most haunting land in Tasmania, the Western Tiers.

After nearly two hours they reached a clearing. Two men and a woman lay wounded on rocky ground around the mangled carcass of a helicopter. A fourth man tried to prop himself up by clinging to a rotor blade. Thoughts flooded Graeme's mind, but they quickly settled as conditioning and training took over. The first task was to make primary assessments of each victim.

All had similar compressed spinal injuries, one also had a ruptured spleen, and the man hanging to the rotor blade was grimacing from a fractured dislocation of the ankle.

The helicopter had crashed several hours earlier. Since then, the pilot had managed to drag himself up a little slope to gain better reception for his emergency phone call, and the man near the rotor blade had hopped and crawled around the others to assist any who needed help. In particular, he attended to the woman, finding as many rugs and clothes as he could to protect her. Then, all of them waited. All they could do was try to keep morale high as the sun went down and the temperature dropped to freezing. By the time the rescue crew arrived, they could only offer weak smiles.

It was impossible to carry all the victims out at once by stretcher, so an aerial rescue seemed the only possible way to deal with the situation. But there had never been a helicopter attempt made from this isolated area, and the difficulties of a night rescue were further complicated by heavy fogs that both sneaked in and left without warning. Nevertheless, an attempt had to be made. After radioing for help, Graeme's colleagues marked a landing zone, piling high clumps of 'kero' bush, so named because it's easy to set alight. And all the while, Graeme moved methodically from patient to patient. The cold bit through his many layers of clothing. His hands were numb, his knees and back ached, but this discomfort wasn't worth a thought when compared with the pain that confronted the people around him. He set up intravenous morphine for each patient, and made them as comfortable as he could. Around him, thin layers of ice formed on the rocks, gusts of wind whipped the trees, and the stars played hide and seek with the fog. Each survivor was in agony, yet they talked calmly to Graeme, showing courage in conditions that could have shattered their spirits. They told Graeme of their lives and their interests. The woman, Pat Frost, was a local councillor, the two other passengers Kelvin Howe and David Kelly, were farmers. As members of the Wildlife and Mountain Hut Preservation Society, they'd been in the helicopter taking materials to a restoration site. When the chopper crashed at 3.10 pm, their understanding of preservation changed dramatically.

At about 2 am, some eleven hours after the accident, the thumping sounds of another chopper cut through the night. Miraculously, the fog cleared away, revealing bright stacks of flame that helped guide the pilot to the ground. Only two patients could be flown out at a time. By 4.30 am, all four had been lifted out. As he listened to the early-morning return to stillness, Graeme quietly thought to himself, 'What a fabulous result.'

After a few hours' sleep, he and the others in the rescue crew walked out the same way they had come in. They left behind a harsh, inhospitable mountainside near Lake Nameless, a place whose anonymity would never be forgotten by both the rescuers and the survivors.

In recalling this incident Graeme deflects attention away from himself. He acknowledges the 'magnificent professionalism and dedication' of all the rescue crew, and in a telling reflection of his character, he has only the highest admiration for those he treated.

'The victims obviously felt as though they owed us gratitude, but they paid us back very much by the way they performed and behaved. They showed tremendous bravery that night. I learnt from them as much as I helped them. I'm very proud of all our work that night. We were orderly, quiet, analytical, and practically sound. Those are the incidents that reflect what being an ambulance officer, or any emergency services person, is all about. It was a very satisfying evening, and one of the most professional jobs I've ever been involved in.'

* * *

Graeme Jones is the ultimate professional, all the way to the clean buttons and neat creases of his uniform. He is also a devoted family man. He and his wife Julie have two children, Sophie and Lachlan, who in years to come will be given the medals that their father received. In his understated way Graeme admits that day will make him very happy because 'when I pass on, my children will be able to say: "my Dad was a good ambulance officer".'

Postscript

Soon after this story was written, Graeme Jones was among a group of paramedics and other medical staff who were involved in the miraculous rescue of Todd Russell and Brant Webb from the Beaconsfield mine about 40 kilometres north-west of Launceston. The two miners spent fourteen nights trapped 925 metres underground after a rock fall on Anzac Day 2006. They were protected from their unstable surrounds by a cage that was part of the telehandler vehicle in which they'd been working at the time of the rockfall. The driver, Larry Knight, was killed. Throughout the long and dangerous rescue Graeme and his colleagues spent many hours underground talking to the trapped miners on a telephone that was threaded through the rock along a PVC pipe that lined a thin eighteen metre-long drill hole. Conversations varied from: recommended exercises that the miners could perform to answering medical questionnaires, from football results to nicknames, anything that came to mind. Apart from providing their own feedback and passing urine samples through the pipe, the miners were observed via a lipstick camera. As the mine rescue team methodically worked through the rock with equipment, explosives and bravery, the seemingly endless hours meant the medical team wasn't only observing two patients, but the patience of everyone involved. After the miners were first found, they were warned not to expect an overnight rescue. They were told they had to treat the incident as a marathon such was the hardness of the rock and the risk of further collapses.

When the miners were finally freed Graeme attended to Todd Russell, who, like Webb, was in remarkably good health and high spirits despite the ordeal. The miners showered and underwent tests in a crib room before they took a lift to the surface and walked with arms raised to their waiting families and mates, their lives changed forever. Graeme was always nearby his patient, and was in the back of one of the two ambulances that pushed through a cheering throng to Launceston General Hospital. They were stirring scenes that inspired a nation.

Much has already been written about this now legendary 'Great Escape', and the names Russell and Webb will never be forgotten. And nor should the deeds of the people whose determination, expertise and courage saved the two mens' lives. Graeme Jones played his role as one of eight ambulance officers and rescue paramedics. He wouldn't have it any other way.

THE ANGELS OF AVOCA

'Community Spirit'

> 'Since it's just been us seven girls I can honestly say that we've never had a fight or an argument in the group. Think about how unreal that is for seven women!'
>
> Volunteer Ambulance Officer Shirley Squires, Tasmanian Ambulance Service

It's 1.48 pm, Thursday 7 July 2005. I approach the door at the end of a long, green tin garage. This is the final stop of my trip throughout Tasmania interviewing ambulance officers and paramedics. I have no idea what to expect; the Chief Executive of the Tasmanian Ambulance Service, Grant Lennox, had given me only one piece of advice before I'd left Hobart: 'Be prepared,' he'd said.

'Prepared for what?'

'Anything.'

I knock. A chorus of crackles answers: 'Come in!'

I enter. The room has two filing cabinets, a desk and computer, a rectangular table that is surrounded by a handful of bright orange vinyl chairs, and a digital clock that squats on a windowsill. On the rear wall, there is a whiteboard covered with indecipherable words written in blue Texta, while the wall to my right boasts a neat row of awards and certificates. But the most noticeable feature is the semicircle of standing women, dressed in a motley range of flannelette jackets, shirts, woollen jumpers, jeans, dresses, trousers and boots. One of them, short and strong, with grey curly hair, extends her hand.

'Hello, I'm Shirley. Thanks for ringing to tell us you were running late. That was really good of you. Come and meet everyone.'

Introductions are made one at a time. There's Mary Knowles, 53, a grandmother. She has been a volunteer ambulance officer for sixteen years. She first joined to learn skills that could one day 'help my husband and children if they got into trouble'. Four of her children were students of distance education where the classroom was the home, and the playground a deer farm.

Fifty-six-year-old Margaret Dennis is the postmistress at the nearby village of Rossarden, once famous for being the largest producer of tin in Australia; the last of the mines was shut in the 1980s. Margaret has lived in Australia since 1973, but her accent can't hide her English origins. She started as a volunteer in the same year as Mary after 'My old man coot (the same sound as in foot) his finger and fainted and I went *Aaaaaaaaaah!*'

Helen Johnson is 57. She is a great-grandmother who has spent all her life in the area. She joined the ambulance service in 1981 as 'it was a way of giving something back for what the community had given me over the years'.

At 59, Helen Reynolds is the 'old duck of the lot of us!' She has a mischievous glint in her eyes, and is also a great-grandmother. She became an officer fourteen years ago 'because

I love to help people, I really do. If I can get in and help someone, then I'll do it.'

Sandi-Lee Squires, 33, a teacher's aide, adds to Helen's explanation with a smirk: 'And don't forget I actually think I talked you into it too.' Sandi-Lee's journey began fifteen years ago when 'I sat in on an ambulance first aid course for a weekend just to see what happened. I ended up doing the exams although I didn't really plan to, and after I passed I just carried on from there. Plus, because of Mum, it was always in the family anyway.' Mum is 56-year-old Shirley, who jumps in off the back of Sandi-Lee's sentence: 'Of course, there's one missing. Jackie can't come because she's looking after the shop.'

Jackie is 30, mother of two, daughter of Shirley, sister of Sandi-Lee, and the youngest member of the group that is the Avoca Volunteer Ambulance Unit.

Avoca is a town of just a few hundred residents in the lap of the rugged Ben Lomond Mountain Range in Tasmania's north-east. It was settled by Europeans in the 1820s, but it wasn't until the following decade that the infrastructure developed after the establishment of a convict probation station, one of several in the region. A sandstone reminder of that period, the Anglican Church of Saint Thomas, still stands in a commanding spot on a hill that overlooks the town and the St Paul's River on its outskirts. Nearby, weatherboard and stone cottages huddle together along the streets, but within a few blocks they make way for sheep pastures, and paddocks in which wheat, triticale and potatoes are grown.

'It's a pretty spot,' says Shirley. 'But when you're isolated, it's got its problems too. That's why we got the ambulance up and going because the community didn't have a doctor that was close. Our nearest one was Campbelltown about 36 kilometres away, and then there was one at St Mary's, about

50 kilometres away. But they were always busy, so we needed to do something.'

Shirley is the longest-serving ambulance volunteer in this group. She joined Avoca's independent service in 1974 as a driver after the townspeople pitched in to buy an old ambulance, a Holden panelvan, from the Aberfoyle mine near Rossarden. A station was established soon afterwards. It was a garage that was just big enough to shelter the ambulance but nothing else. When a new vehicle, a Ford F100 was acquired, Shirley recalls there was need for considerable change: 'When we got it [the Ford], it wouldn't fit into the shed that'd been built for the Holden. So we built another shed. It had a roller door on it, and we were all happy with it, but one day one of our officers forgets to push the door right up and he ends up with two lights missing off the top of the vehicle. He'd chopped 'em clean off!'

During the early years Shirley did regular training courses at St Mary's Hospital, and once a fortnight she travelled the 80 kilometres to Launceston for more intensive lessons. The independent service survived for a number of years until the cost of running it far exceeded the money gained by subscription, chook raffles and government support.

In 1990, the Tasmanian Ambulance Service assumed control. Several years later, a new station was built adjoining the fire station. It exists to this day, and is where laughter now fills the room as I listen to a group of cheerful women, who, despite the traumas they've experienced, know how to treat life with a smile. They're known as the Avoca Angels, a tag that brought grins all round not too long ago when a tall, handsome 20-year-old shearer, Aaron Milner, was the eighth member of the unit. He was nicknamed 'Bosley' in reference to the television character of the same name in the 'Charlie's Angels' series.

Aaron has since moved on, leaving his ladies to follow scripts that Hollywood wouldn't dare pursue.

'We've been called the "Angels of Mercy" in the paper, you know', says Helen Reynolds leaning forward in her seat.

There is more laughter.

She continues: 'It's just something we do. The community sticks together, and we work together. I'd miss it if I stopped because it has become part of me.'

'Yes and you get better with age!' adds Shirley.

The banter in the room is beautiful. Here are six women who not only enjoy but cherish each other's company.

'So all you want us to do is talk?' asks Shirley.

'Oh, that's easy,' replies Helen Reynolds. 'It's stopping us that's hard!'

And so they start telling their stories. Shirley begins, periodically swallowing her chuckles as she recalls the day in Launceston when a 'huge bloody dog as big as a table' chased her down the street and into the back of the ambulance. Then there was the time that a patient fell off a chair being lifted by two officers; it prompted a chain reaction that ended with all six people sprawled on the ground legs up in the air.

'And I also remember a car accident we went to on a really cold day,' says Shirley. 'It was as frosty as hell . . . Oh no, hell's hot isn't it? Oh well, it was frosty anyway.'

Helen Johnson breaks in to the conversation. Until now she has been sitting quietly while chewing her bottom lip. She unfolds her arms and begins: 'One of my cases involved one of my own grandchildren. He was only a baby. He got scalded. I got there and our ambulance was already out on another job. Basically we had to get him to help on the back seat of a passenger car. We had to get him to the doctor, so a phone call is put through to Campbelltown ambulance, and we say we'll meet them at the Conara turn-off (about 25 kilometres away). They turned up and I helped them with the treatment. After I got back into my son-in-law's car, I just couldn't do anything. There was no way I was able to move, I couldn't even put a cigarette in me mouth, me hands just wouldn't do anything I wanted them to. Me legs were like jelly. It's different when it's your own. If you talk to me about being on the job it's fine, but afterwards, it can take time.'

There is a brief moment of silence, long enough for everyone to consider the significance of what Helen has said. Then, she continues as though she has sensed a need to change beat: 'You also get the other side, you know. Margaret and I went to this patient once; he'd been mowing the lawn and a rock had hit him and cut him. His blood pressure had gone through the floor. You couldn't get it; there was no reading whatsoever when we first got to him, but it slowly started to come back. We put him in the ambulance and headed to Campbelltown. The paramedic from there was on his way out and he met us along the way. He had a look at the patient and told us we should take him into Campbelltown, and wait for him to stabilise and then we could bring him home again. So we went and did that. Well, the poor bloke [the patient] felt so embarrassed and disgusted because he felt as though he'd wasted our time. He was so downhearted. As it turned out, on the way back home we pulled up the first on the scene of an accident. There's this other roving paramedic who's been out in the country, and he's come out in his station wagon on his own. Well, he knew the victim needed treatment, so we put him in the back of the ambulance, and Margaret drives this paramedic's vehicle while the paramedic works on the patient. But you see I can't drive, so we have to ask the lawn-mower patient to drive the ambulance. So he turns it around and we start heading back to Campbelltown. Then the paramedic that we'd seen earlier comes toddling along and sees us, and you can see him thinking: "Isn't the bloke driving meant to be a patient?" Well, we sort it all out. The roving paramedic gets back in his vehicle and goes off to where he was going, the Campbelltown paramedic takes the [accident] patient and goes off to Launceston, and we go home. And all the way the lawn-mower man is much better because he feels as though he hasn't wasted anybody's time. We dropped him off and he was feeling really good about it all.'

Laughter bounces all over the room as the stories continue. Sandi-Lee recalls when she and Helen Reynolds bogged the ambulance, and the other occasion when they were asked by Central Radio Communications where they were. Helen simply replied: 'Where the cows are,' in reference to a nearby feedlot.

As is the way of this conversation, the topic quickly changes again when Helen Reynolds notes that a picture of the 'Angels' was in the regional paper, *The Examiner*, earlier in the week: 'I walked through the park this morning and I see someone who says: "Quick, where's a piece of paper?" 'Cos you see they wanted my autograph. And I shook my head and told them: "Don't even start!" A friend of mine had rung me up when we were put on the front cover of the phone book, and you should've heard him! So I thought "I'll fix you, mate!" So I sign this piece of paper and send it to him. Well, he framed it didn't he?'

There is an explosion of laughter, after which Mary pounces: 'We'd organised an extra UHF radio to pick up for the farm from this fella at Launceston. When I was ringing him the night before I asked him to hold it for me, and if he needed to check my credentials he could just look at the photo in the paper. Anyway when I get out there, he comes over to the car and says: "Oh Mary, I'm so pleased to meet you, you're so wonderful, you people do such a good job. I could have sold this radio 40 times over but I kept it for you."'

Helen Reynolds chips in: 'You know how they have those things at school about what you would like to be when you grow up? Well my grand-daughter said she would like to be an ambulance officer just like her Nan.'

'Does that mean she's gonna break everything?' asks Sandi-Lee.

'Yes!' says Shirley, grinning. 'We call Helen "Wrecker Reynolds" because whatever she puts her hand on she wrecks!'

'I didn't break anything until Helen joined,' adds Sandi-Lee.

Amid the merriment, Shirley finds an irresistible opening: 'We got a call out one day to go to Rossarden,' she says.

'Oh, no, don't put that in there!' replies her daughter.

Shirley gleefully takes no notice, folding her hands across her lap, and resting deep into her chair: 'We've got a battery warmer plugged into the ceiling out here in the shed. Sandi-Lee gets in, she lifts the door up, jumps in the bus [ambulance] and whizzes off. She says, "We don't seem to be pulling right," and I replied, "Well, just check the back window, it could be the whole shed behind us!"'

She slaps her thigh, and predictably the room erupts. Any person hearing the noise from the street outside would surely want to come in and join the apparent party, but just as it seems there is no limit to this regalement, Shirley pauses, squints through her glasses, and frowns. The tone of her voice drops after she takes a long breath.

'You know we've had some good accidents, and we've had some bad ones. Ah the girls are probably sick of me saying this, but you always have one that really sticks with you.'

No one says a word. Shirley continues: 'One night we got a call-out to a motor vehicle accident, and there was a head-on collision, and it was out along the Mathinna road. When we were on the way there, we heard there were fatalities, there were three of them. And when I got out there, they were all my family. Nieces and nephews . . .'

Shirley begins to cry. She bites her lip, stands, and walks out of the room. The door squeaks open and bangs shut behind her.

'I think it's time for a cuppa', says Mary.

Everyone agrees.

After slices of lamington roll and a strawberry sponge cake are washed down with 'white and ones' or 'black and none', everyone, including Shirley, settles down for more chat.

The women resume with a discussion about the gratitude of the community. Sometimes they receive thank you phone calls, cards or letters, and it has been known for flowers and boxes of chocolates to arrive on their doorsteps. Shirley remembers the day she attended a boy who'd fallen off his motorbike; that night she answered a phone call from the boy's mother who suggested her son 'couldn't have been looked after better even if it was by his own mum'. Shirley smiles.

They all agree that they're relied on heavily, perhaps too much so on the occasions when someone knocks on their door and asks for a Bandaid for a cut finger, or medicine for a cold. The Angels aren't considered by some to be volunteer ambulance personnel, but doctors who will attend every need. But there is also the flipside where courtesy overrides commonsense; people, particularly the elderly, will choose to suffer through the night because they don't want to be rude and wake anyone.

From house calls to car accidents, the removal of stitches (something they no longer do) to administering advice as medicine, the Angels have been there at some stage for a considerable portion of their community. But with that come the difficulties of providing a service in an area where your next-door neighbour could write your life story. Sandi-Lee explains: 'You go out on a job, and nine times out of ten, there are people in town who already know where you're going. And by the time you come back, someone will know all the details. They'll give you all the ins and outs as if they were there themselves. But then there are other times when you come back and everyone wants to know every detail. That can be really hard as everything is confidential. We had a case a few years ago when we got a call-out while I was at school. The police needed us to verify that a poor old guy had died. When we got to his home the poor devil was sitting in his chair and he was deceased, so we came back, and I went back to school. And of course every time I come back the children are always asking: "Where were you?" "What happened?" "Are they alright?" You get 100 questions. On this particular day, the grandson of the dead man happened to come up to me and ask: "How are they?" And all I could say was: "Much the same as when we left him". Of course I couldn't say anything, knowing that this boy and his sister were going to go home that afternoon and have a horrible time.'

In such cases, the Angels take strength from each other. Margaret, the quietest of the group, acknowledges everyone works well together 'partly because we all come to training every second Monday; it's rare for anyone to miss it.' Shirley adds: 'Because we've all done the training we can all do the job. No one is better than the next person.' Then Sandi-Lee jumps in: 'And we all still have the same "Oh Shit Factor!"'

There is another round of laughter, much quieter than earlier bursts, and much more poignant. Helen Johnson shoulders it aside: 'We also have something else. When we go out, regardless of what we do, especially if it's a bad job, we've got each other to talk to coming home. And that's the most important thing.'

Every head in the room quietly nods before Shirley adds 'And you have to laugh. My motto is that you have to have a weird sense of humour to be here. A little bit of laughter can be the best medicine for you and your patients. And it doesn't cost you nothing. Personality and attitude go a long, long way, and that's what this group has got.'

'Oh yeah,' replies Helen Reynolds, nearly falling off the edge of her seat with enthusiasm, and looking at Sandi-Lee as though giving the signal for a double-act. 'What about that bloke from Alaska. He'd never seen a snake in his life. Him and his wife. Never ever.'

'It went across the road in front of him,' adds Sandi-Lee, picking up the lead.

'No, it was dead on the road.'

'Anyhow, whatever it was doing, it made the bloke do a U-turn on a blind corner to go back and have a look, and he caused a three-car pile-up. It was only a tiny car he had, but he basically wrote off everyone else's vehicle.'

It is now approaching 4 pm, and silly hour appears to be arriving as Shirley unveils one of her favourites: 'This has nothing to do with our group, but it's worth telling. We went to training up at St Mary's about three Sundays ago and heard this one. A paramedic was a few minutes behind the ambulance, and the officer in the ambulance radioed through to Comms [Central Communications] and said: "Could you get in contact with the paramedic behind us and tell him to be careful because there's a black cow on the road."

'And Comms came back and say: "Aah, you did say a black car?"

'"No," says the ambulance officer. "I said a black cow."

'And Comms asked again, "Did you say a black car?"

'"No," says the ambulance officer. "I said a black *MOOOOOOOOOOOOOO!*"'

The loudest burst of laughter for the afternoon bounces off every wall and brightens every face.

It is impossible to predict what turn the conversation will take next. Seriousness offers a guiding hand for a while as the Angels discuss the state-wide Volunteer Ambulance Officers' Association of which Mary is president; Margaret secretary; and Shirley, treasurer. It seems that the more these women help the community, the more they feel the need to keep giving.

Sandi-Lee, the third officer of the Avoca Bush Fire Brigade, and a volunteer at Camp Quality (a non-profit organisation that provides recreational, social, and financial support for children suffering from cancer) acknowledges: 'There are always pies; it's just a matter of finding enough fingers!' She pauses, a warm smile curls into her cheeks.

Mary suddenly pipes up: 'Listen, I've got a funny story for you. Helen [Reynolds] and I were going to town once. We were going along the Midlands Highway towards Launceston, and Helen just stops, ducks off the road, jumps out of the ambulance and runs! And I thought, "Well what the hell is going on here?" I've got a patient in the back with broken ribs that we've gotta get into town, and there's Helen heading for the hills. So I climb out and go after her and ask her what's going on. And she says: "There's a trodgy in there. I am not going back in that car with a trodgy in it!" So I go back to the ambulance, grab a towel, and you could see this trodgy's legs on the sun visor. So I just get him down, shut me eyes, grab the damn thing, throw the towel outside and say to Helen "Get in, we're going!"'

The hysteria all but engulfs Mary's explanation that a 'trodgy' is a huntsman, the hairy long-legged spider that has lifted many a human heart rate into orbit. Helen Reynolds shakes her head, and the creases in her forehead contort: 'That bloody trodgy had longer legs than me! I wasn't going near it! No bloody way!' Then she leans back and laughs.

She laughs hard.

Everyone does.

'You see,' says Shirley. 'I told you laughter is the best medicine.'

The room falls quiet. Perhaps the Angels have exhausted themselves or, more likely, they're just taking a moment or two to catch their breaths. Just as they have done in their ambulances for many years, they have, this afternoon, covered many miles. The red numbers on the digital clock on the windowsill click over to 4.40 pm. Three hours have passed since I arrived, and now it's time to leave, but I still have one more question: What does being a volunteer ambulance officer mean to each of you?

Silence follows, then one at a time they answer.

Helen Reynolds: 'It's about giving time. My husband asks me when he can make an appointment to see me. There's this house here that has a sign out the front "Seldom Inn". My husband reckons he's gonna make one for our house: "Never Inn!" So, it's all about making time, and giving it. It's not hard, you know.'

Helen Johnson: 'Helping others. It's about being there when someone needs you. And when you help others, you know you can also help yourself too.'

Margaret: 'It's being part of the community. In small places like here we need each other. We have to be there for each other.'

Mary: 'I've got more out of it than what I've put into it. It really comes down to knowing what to do. I'm glad I do. I am better off for it, and I know others are too.'

Sandi-Lee: 'It's like what everyone else says. It's giving yourself and your time freely to help others. It's not all lights and sirens as some people might think. It is a job where there isn't always thanks, but the look of relief on the faces of patients, family and friends when we arrive is thanks enough. That is reason enough, and the reason why I keep doing this job. I'm sure this would be similar for all who volunteer, regardless of the job.'

Shirley is the last to answer. Thirty-one years of experience guides her thoughts. She speaks firmly, speaks with authority: 'First, let me say that because of what we've done over the years we'd be lost if we had to give up volunteering tomorrow; we wouldn't know what to do with ourselves. We are all proud to be members of the ambulance service, proud to wear our uniform, and be able to serve the community and our fellow human beings.'

Silence soaks up these words. Little else needs to be said, except my request for a photo. Without any encouragement these wonderful women change into their blue uniforms and stand proudly. Click! Everyone smiles: Mary Knowles; Margaret Dennis; Helen Reynolds; Helen Johnson; Sandi-Lee Squires and Shirley Squires. And let's spare a thought for Jackie Squires who is too busy running the town store to pause for conversation and a camera.

Our farewells begin with handshakes and end with hugs. I leave with a grin, and a copy of *Rossarden Recipes*, a book that some of the Angels were instrumental in producing; it's not the least surprising to be told that one dollar from each book sold goes to the Fred Hollows Foundation, with the remaining proceeds funding a Christmas party for the kids of Rossarden.

The goodbyes are over. I have one foot out the door, and my hand is reaching into a pocket for my car keys when there is a familiar chirp behind me. I turn to see Helen Reynolds standing in the middle of the group, waving her finger at me: 'Now you take care driving back. We loved talking to you but we don't want to see you again today!'

I smile, shut the door and walk away from the long green tin garage that houses a vehicle in which lives are saved and lost. It is driven by the most powerful force in the world... human spirit.

Author's note: After this story was written I was contacted by Jackie Squires (the one missing Angel on the day I visited.) She asked if she too could express why she became an Angel. This is what she told me:

'I never liked blood at all, then one night I was at a cabaret with Mum when a man was assaulted. I helped her treat his injuries before the ambulance arrived. I thought "that wasn't so bad" (of course there was blood involved). So I thought I could give a bit back to the community by helping others and myself at the same time, especially since I have two young boys.'

Thank you, Jackie. Thank you, Angels!

EPILOGUE

I would never have had the opportunity to write this book if the former publishing director of Hachette Livre Australia, Lisa Highton, hadn't been in a coffee shop one morning when two ambulance officers walked in. She watched them order their coffee and muffins, and then they merrily went on their way for the day. It made Lisa think about what was ahead for them. She went to her office, the ambos went to the front line, and the idea for *Just Doing My Job* went to the discussion table.

When I started this project I knew only a few police officers, one or two firefighters, and not one ambulance officer. Nor did I have anything more than a basic understanding of their roles. However, over the following nine months both my address book and knowledge grew substantially. I had a range of both awful and wonderful experiences that gave me a brief glimpse into life at the front line. These included attending the aftermath of a suicide; sweeping rising floodwaters away from houses; riding in the back of fire trucks roaring through city streets;

having Sunday roasts and lamington rolls; crying, laughing, hugging, shaking hands . . .

Most importantly, I met some wonderful people. They are everyday people who you could walk past in the street without ever noticing. They are not superstars, and certainly don't wish to be in the spotlight for their deeds: modesty is one of their strongest traits. After completing each interview for this book, I departed with a greater belief in the human spirit. I was very, very lucky to have had this opportunity and we, the general public, are very, very lucky to be served by people who are driven by 'just doing their jobs'.

ACKNOWLEDGEMENTS

Researching, writing, and producing *Just Doing My Job* required an immense team effort.

Firstly, thank you to the subjects whose enthusiasm, professionalism, cheerfulness, and interest made this all possible: Iain Argus, Bernie Aust, Julian Bisbal, Kevin 'Billy' Boyle, Mark Burns, Louise Cannon, Danny Carson, Rowan Chapple, Dave Cuskelly, Margaret Dennis, Anya Dobaj, Julie Elliott, Peter Gill, Darren Hay, Colin and Robyn Heterick, Helen Johnson, Graeme Jones, Dave Kelly, Mary Knowles, Justin Morrison, Eric Noble, Andrew O'Connell, Garth Pitman, Grant Pitman, Kendall Pitman, Stacey Pitman, Greg Plier, Helen Reynolds, Ivy Rooks, Kerrie Shepperd, Craig Simpson, Neale Smith, Dave Stevenson, Pat Stevenson, Gary Squires, Sandi-Lee Squires, Shirley Squires, Shane Torr, Stephane Victor and Deb Wallace. I will be forever grateful for your tremendous feedback and support, and I look forward to sharing the occasional drink and yarn with you as the years go by. Furthermore, a very special thank you to all the people who helped me put together the story about the late Mark Mansfield: Debra Mansfield, Rachael Mansfield, Doreen Mansfield, Sarah Mansfield, Matthew Mansfield, Christine Carr, Lyn Adams, Donald Adams, Peter Morgan, Phil Brumby, Christine Holden.

I am also indebted to the organisations that offered their enthusiastic support for this book:

Queensland Police Service (www.police.qld.gov.au)
Queensland Ambulance Service (www.ambulance.qld.gov.au)
Queensland Fire and Rescue Service (www.fire.qld.gov.au)
NSW Police Service (www.police.nsw.gov.au)
ACT Fire Brigade (www.firebrigade.act.gov.au)
Victoria Police (www.police.vic.gov.au)
Metropolitan Fire and Emergency Services Board
(www.mfbb.vic.gov.au)
Rural Ambulance Victoria (www.rav.vic.gov.au)
Tasmanian Ambulance Service (www.dhhs.tas.gov.au)
South Australian Metropolitan Fire Service
(www.samfs.sa.gov.au)
SA Ambulance (www.saambulance.com.au)

Thank you to the staff from the above organisations whose professionalism and friendliness made my job so much easier and so very enjoyable. I particularly extend my gratitude to Sonia Giovannetti (Queensland Police Service Media); Sarah McCormack (Queensland Fire and Rescue Service Media and Public Relations); Ben Creagh (Queensland Ambulance Service Media and Public Relations); Kate Longton (NSW Police Service Media); Darren Cutrupi (ACT Fire Brigade Media and Public Relations); Jessica Li (Rural Ambulance Victoria Media and Public Relations); Constable Stephen Gambetta (Victoria Police Film and Television Office); Angelique Pantazi (Metropolitan

Fire and Emergency Services Board Public Relations and Events); Grant Lennox (Chief Executive Officer, Tasmanian Ambulance Service); Greg Crossman (Command District Officer, South Australian Metropolitan Fire Service); John Foody (South Australian Metropolitan Fire Service Media and Public Relations); and Lee Francis (South Australian Ambulance Service Emergency and Major Events).

There were numerous others who helped during the interview and writing stages whom I must also thank: Chris Bitter, Linda Boyle, Mrs Chatterton, Bob Fraser, Megan Gavel, Dr John Greenwood, John Harris, Amy Li, Cameron McKenzie, Kirsten Miller, Denise Morcombe, Sarah Morrison, Bobby Nairn, Chris Nottle, 'Old Ray', Tjamme Rooks, Gary Shanks, Collin Ween, Leon Shepley, Steve Wallace-Yarrow, and Graham Wood.

Thank you too to the great staff at Hachette Livre. To Lisa Highton, who has now returned to her homeland, England. Lisa, I am very grateful that you had faith in me, and I hope these stories uphold your original vision for the project. As with every book, there is a large production team that works behind the scenes that deserves great credit: Commissioning Editor Vanessa Radnidge (a very big hug to you, Vee!); Anna Waddington, Janis Barbi, Fiona Hazard, Emma Rusher, Lisa Hunter, Matt Richell, Luke Causby and Simon Paterson.

I also thank my close friends and family who put up with my stubbornness and single-mindedness from the moment I started *Just Doing My Job*. In particular, the very biggest of hugs to my agent Jane Burridge, Richard and Carissa Tombs, Phil Gregson, Iain Bellamy, Tabitha Pearson, Peter Overton, and my mother Anne. Mum, you will always be the greatest teacher I've ever had. I love you. Finally, I couldn't have done this without the love and support of my wife Clare. Your warmth, kindness, calmness and bouts of utter silliness remind me every single day why I'm the luckiest bloke on the planet.

Thank you all. It has been a wonderful experience.

Best-selling author James Knight was raised in the New South Wales country town of Gunnedah. In a media career spanning nearly twenty years he has been a reporter and producer for television channels Seven, Nine, Ten and Fox Sports, and he is a freelance documentary maker. He has been a newsreader for Sydney radio stations 2SM and 2DU, and a writer for *The Sydney Morning Herald*, numerous magazines and advertising houses. He has also worked in India as a media consultant. He is a popular motivational speaker who draws strength and inspiration from the many people he meets. When not working he is an intrepid traveller who enjoys running marathons, and sharing new experiences and adventures with his wife Clare. His motto is: 'Try the impossible because you just might succeed.'

Just Doing My Job is James Knight's fourth book. It follows: *Lee2*; *Mark Waugh: the Official Biography*; and *The Dragon's Journey*.

www.knightwriter.com.au